AF846063

The Role of Experimentation in Building Future Naval Forces

Committee for the Role of Experimentation in Building Future Naval Forces
Naval Studies Board
Division on Engineering and Physical Sciences

NATIONAL RESEARCH COUNCIL
OF THE NATIONAL ACADEMIES

THE NATIONAL ACADEMIES PRESS
Washington, D.C.
www.nap.edu

THE NATIONAL ACADEMIES PRESS 500 Fifth Street, N.W. Washington, DC 20001

NOTICE: The project that is the subject of this report was approved by the Governing Board of the National Research Council, whose members are drawn from the councils of the National Academy of Sciences, the National Academy of Engineering, and the Institute of Medicine. The members of the committee responsible for the report were chosen for their special competences and with regard for appropriate balance.

This study was supported by Contract No. N00014-00-G-0230, DO #12, between the National Academy of Sciences and the Department of the Navy. Any opinions, findings, conclusions, or recommendations expressed in this publication are those of the author(s) and do not necessarily reflect the views of the organizations or agencies that provided support for the project.

International Standard Book Number 0-309-08873-9 (Book)
International Standard Book Number 0-309-50851-7 (PDF)

Copies available from:

Naval Studies Board
The Keck Center of the National Academies
500 Fifth Street, N.W., Room WS904
Washington, DC 20001

Additional copies of this report are available from National Academies Press, 500 Fifth Street, N.W., Lockbox 285, Washington, DC 20055; (800) 624-6242 or (202) 334-3313 (in the Washington metropolitan area); Internet, http://www.nap.edu.

Copyright 2004 by the National Academy of Sciences. All rights reserved.

Printed in the United States of America

THE NATIONAL ACADEMIES

Advisers to the Nation on Science, Engineering, and Medicine

The **National Academy of Sciences** is a private, nonprofit, self-perpetuating society of distinguished scholars engaged in scientific and engineering research, dedicated to the furtherance of science and technology and to their use for the general welfare. Upon the authority of the charter granted to it by the Congress in 1863, the Academy has a mandate that requires it to advise the federal government on scientific and technical matters. Dr. Bruce M. Alberts is president of the National Academy of Sciences.

The **National Academy of Engineering** was established in 1964, under the charter of the National Academy of Sciences, as a parallel organization of outstanding engineers. It is autonomous in its administration and in the selection of its members, sharing with the National Academy of Sciences the responsibility for advising the federal government. The National Academy of Engineering also sponsors engineering programs aimed at meeting national needs, encourages education and research, and recognizes the superior achievements of engineers. Dr. Wm. A. Wulf is president of the National Academy of Engineering.

The **Institute of Medicine** was established in 1970 by the National Academy of Sciences to secure the services of eminent members of appropriate professions in the examination of policy matters pertaining to the health of the public. The Institute acts under the responsibility given to the National Academy of Sciences by its congressional charter to be an adviser to the federal government and, upon its own initiative, to identify issues of medical care, research, and education. Dr. Harvey V. Fineberg is president of the Institute of Medicine.

The **National Research Council** was organized by the National Academy of Sciences in 1916 to associate the broad community of science and technology with the Academy's purposes of furthering knowledge and advising the federal government. Functioning in accordance with general policies determined by the Academy, the Council has become the principal operating agency of both the National Academy of Sciences and the National Academy of Engineering in providing services to the government, the public, and the scientific and engineering communities. The Council is administered jointly by both Academies and the Institute of Medicine. Dr. Bruce M. Alberts and Dr. Wm. A. Wulf are chair and vice chair, respectively, of the National Research Council.

www.national-academies.org

COMMITTEE FOR THE ROLE OF EXPERIMENTATION IN BUILDING FUTURE NAVAL FORCES

ANNETTE J. KRYGIEL, Great Falls, Virginia, *Chair*
RUZENA K. BAJCSY, Center for Information Technology Research in the Interest of Society, University of California, Berkeley
ALAN BERMAN, Applied Research Laboratory, Pennsylvania State University
DUNCAN A. BROWN, Applied Physics Laboratory, Johns Hopkins University
MARION R. BRYSON, North Tree Management
JOHN D. CHRISTIE, Logistics Management Institute
JOHN A. CORDER, Colleyville, Texas
PAUL K. DAVIS, RAND and RAND Graduate School of Policy Studies
BRIG "CHIP" ELLIOTT, BBN Technologies
J. DEXTER FLETCHER, Institute for Defense Analyses
RICHARD J. IVANETICH, Institute for Defense Analyses
L. DAVID MONTAGUE, Menlo Park, California
WILLIAM B. MORGAN, Rockville, Maryland
JASON PROVIDAKES, MITRE Corporation
JOHN E. RHODES, Balboa, California
WILLIAM D. SMITH, Fayetteville, Pennsylvania
MICHAEL G. SOVEREIGN, Monterey, California
MITZI M. WERTHEIM, Center for Naval Analyses
CINDY WILLIAMS, Massachusetts Institute of Technology
JOSEPH ZEIDNER, Bethesda, Maryland

Staff

RONALD D. TAYLOR, Director (on leave as of July 12, 2003)
CHARLES F. DRAPER, Acting Director (as of July 12, 2003)
MARY G. GORDON, Information Officer
SUSAN G. CAMPBELL, Administrative Assistant
IAN M. CAMERON, Project Assistant
SEYMOUR J. DEITCHMAN, Consultant
SIDNEY G. REED, JR., Consultant
JAMES G. WILSON, Consultant

NAVAL STUDIES BOARD

VINCENT VITTO, Charles S. Draper Laboratory, Inc., *Chair*
JOSEPH B. REAGAN, Saratoga, California, *Vice Chair*
ARTHUR B. BAGGEROER, Massachusetts Institute of Technology
ALAN BERMAN, Applied Research Laboratory, Pennsylvania State University, *Special Advisor*
JAMES P. BROOKS, Litton/Ingalls Shipbuilding, Inc.
JOHN D. CHRISTIE, Logistics Management Institute
RUTH A. DAVID, Analytic Services, Inc.
ANTONIO L. ELIAS, Orbital Sciences Corporation
BRIG "CHIP" ELLIOTT, BBN Technologies
FRANK A. HORRIGAN, Bedford, Massachusetts
JOHN W. HUTCHINSON, Harvard University
RICHARD J. IVANETICH, Institute for Defense Analyses
HARRY W. JENKINS, ITT Industries
MIRIAM E. JOHN, Sandia National Laboratories
DAVID V. KALBAUGH, Applied Physics Laboratory, Johns Hopkins University
ANNETTE J. KRYGIEL, Great Falls, Virginia
L. DAVID MONTAGUE, Menlo Park, California
WILLIAM B. MORGAN, Rockville, Maryland
JOHN H. MOXLEY III, Korn/Ferry International
ROBERT B. OAKLEY, National Defense University
NILS R. SANDELL, JR., ALPHATECH, Inc.
JAMES M. SINNETT, Ballwin, Missouri
WILLIAM D. SMITH, Fayetteville, Pennsylvania
RICHARD L. WADE, Risk Management Sciences
MITZI M. WERTHEIM, Center for Naval Analyses
CINDY WILLIAMS, Massachusetts Institute of Technology

Navy Liaison Representatives

RADM LEWIS W. CRENSHAW, USN, Office of the Chief of Naval Operations, N81 (through June 6, 2003)
RADM JOSEPH A. SESTAK, JR., USN, Office of the Chief of Naval Operations, N81 (as of July 15, 2003)
RADM JAY M. COHEN, USN, Office of the Chief of Naval Operations, N91

Marine Corps Liaison Representative

LTGEN EDWARD HANLON, JR., USMC, Commanding General, Marine Corps Combat Development Command

Staff

RONALD D. TAYLOR, Director (on leave as of July 12, 2003)
CHARLES F. DRAPER, Acting Director (as of July 12, 2003)
MICHAEL L. WILSON, Program Officer
MARY G. GORDON, Information Officer
SUSAN G. CAMPBELL, Administrative Assistant
IAN M. CAMERON, Project Assistant

Preface

As we enter the 21st century, the Department of Defense (DOD) seeks to transform the nation's armed forces to meet the military challenges of the future. The absence of a threatening major power in today's world offers the DOD a rare opportunity to experiment, change, innovate, and transform its forces to meet tomorrow's needs while at the same time addressing today's missions. Various reviews currently under way are seeking to establish strategic guidelines for building tomorrow's joint military forces. In addition, activities such as the DOD's fiscal year 2001 Quadrennial Defense Review help to ensure that Navy and Marine Corps programs, processes, and organizations, and the capabilities that they create, are integral to realizing the objectives of joint forces. In this context, the development of joint warfighting capabilities is among the most important of the future issues facing the Department of Defense (and the Department of the Navy), and the recent war with Iraq has accelerated recognition of future requirements and the development of concepts to address them.[1]

During the past decade, experimentation has taken on increased importance in building naval force capabilities. Through its fleet battle experiments, the Navy has attempted to explore and use emerging systems and technologies in order to develop new operational concepts. The Marine Corps Warfighting

---

[1]The present study concluded at the end of Operation Iraqi Freedom, just as assessments of lessons learned were being initiated. The U.S. Joint Forces Command, for instance, was assigned to meet with assessment teams from all of the Services to collect their respective observations on the war. See Malina Brown, 2003, "Thornberry Questions Services' Objectivity: USJFCOM Team to Meet with Services on Lessons Learned from Iraq War," *Inside the Navy*, Vol. 16, No. 18, May 5, p. 1.

Laboratory has conducted experiments designed in part to identify new operational concepts and the capabilities that would be needed to support such concepts as Expeditionary Maneuver Warfare, Operational Maneuver From the Sea, and Ship to Objective Maneuver. The U.S. Joint Forces Command is now charged with leading the transformation of the armed forces and meeting the national security challenges of the 21st century, in addition to being the primary catalyst for joint force integration, training, experimentation, doctrine development, and testing.

Fertile areas for potential gain and progress are found in all three operational domains—land, sea, and air (including space). For example, the military effectiveness of ground forces (Marine Corps and Army) could be increased and cost savings realized if there were agreement on common requirements for command, control, communications, computers, intelligence, surveillance, and reconnaissance (C4ISR) equipment and products, common and complementary operational concepts, and common training technology. Similar gains could be achieved for air operations through the adoption of common Navy and Air Force approaches to conducting such operations. Used to implement and evaluate the networking of joint forces and the development of joint operational architectures as well as interoperability across the Services' systems, experimentation could play a significant role among all Service components in enhancing naval (and joint) force development.

Indeed, experimentation serves as a critical underpinning for the Navy's strategy of transitioning to a network-centric naval force. It is through well-conceived and well-designed experiments (namely, technical demonstrations) that the naval forces will identify new command relationships for conducting military operations, discover information requirements necessary to support various concepts of operation, and learn how to operate in the face of degraded levels of service when under information attack.

In the near term, experimentation allows for operational and technical improvements to current force capabilities and, in some cases, for additional exercising and training of forces, thereby helping to maintain readiness. Near-term experimentation will also greatly affect long-term force development, by identifying areas in which investment will be necessary to support future operational concepts, as well as by introducing emerging technologies to meet the evolving challenges presented to naval (and joint) warfighters.

TERMS OF REFERENCE

At the request of the Chief of Naval Operations (CNO), the Naval Studies Board of the National Research Council conducted a study to examine the role of experimentation in building future naval forces to operate in the joint environment. The study addresses the opportunities offered by experimentation, the implications of experimentation, and the following questions:

• What has been learned from the experiments thus far about future naval force operations and about joint operations involving the fleet and Marines?

• How has spiral development been involved and has it improved timelines and affordability? How does new technology or equipment improve naval force performance and how is spiral development best implemented to achieve desired objectives?

• How successful has been the transitioning of the results of experimentation to the field?

• How adequate are the tools and environments for experimentation (e.g., Navy's modeling and simulation capabilities, integration facilities, etc.)?

• What important questions remain or have been raised or are being raised that are amenable to experimentation and that may or may not be on the agenda? What should be added, and how should such questions be approached (battle laboratories, fleet experiments, joint experiments, and such)?

• What process and method improvements for coherent experiment planning are needed?

• How can and should joint experimentation leverage Service experiments? How successful have Navy and Marine Corps experimentation programs been at preparing for joint operations?

COMMITTEE MEETINGS

• *April 4-5, 2002, in Washington, D.C.* Organizational meeting: Office of the Chief of Naval Operations (OPNAV), N70, overview of naval experimentation and the Naval Transformation Plan; U.S. Joint Forces Command (USJFCOM), J9, briefing on experimentation organization and management; Office of Naval Research (ONR) briefing on naval science and technology programs, implementation, progress, and examples; Marine Corps Warfighting Laboratory overview of experimentation efforts; Navy Warfare Development Command (NWDC) overview of U.S. Navy experimentation and experimentation organization; and Office of the Assistant Secretary of Defense for Command, Control, Communications, and Intelligence overview of previous studies and reviews covering experimentation.

• *May 2-3, 2002, in Washington, D.C.* Office of the Chief of Naval Operations, N79, briefing on Navy education and training in the context of experimentation; Marine Corps Systems Command (SYSCOM) briefing on Marine Corps systems engineering efforts and the SYSCOM's role in spiral development; Marine Corps Combat Development Command briefing on Marine Corps concepts and doctrine development and education and on training efforts; Office of Naval Research briefing on extending the littoral battlespace; Naval Sea Systems Command (NAVSEA) briefing on the Navy's systems engineering efforts and the SYSCOM's role in spiral development; Office of the Assistant Secretary of the Navy for Research, Development, and Acquisition (ASN (RDA)) and NAVSEA briefings on the status of ASN (RDA) chief engineer; and retired leaders' perspectives on Army and Navy experimentation.

- *July 9-11, 2002, in Newport, Rhode Island.* Site visit to NWDC for an overview of NWDC and briefings from NWDC Concept Development Department, Technology Department, Operations Department, Modeling and Simulation Department, Doctrine Department, and the Maritime Battle Center, as well as an overview of NWDC Assured Access Warfighter Innovation Development Team (WIDT), Forward Sea-Based Forces WIDT, Anti-Terrorism/Force Protection WIDT, Information Knowledge Advantage WIDT, and Effects Based Operations WIDT; and CNO Strategic Studies Group briefing on FORCEnet.
- *July 30-August 1, 2002, in Hampton Roads, Virginia.* Site visit to USJFCOM for an overview of USJFCOM activities, past and present, and briefings on Millennium Challenge '02; video discussion with Commander, Second Fleet, U.S. Navy, on the fleet perspective on naval and joint experimentation; Joint Warfare Fighting Center briefings and demonstrations of the ongoing Millennium Challenge '02; Submarine Force, U.S. Atlantic Fleet and Submarine Development Squadron 12 overview and perspective on experimentation; U.S. Atlantic Fleet perspective on naval and joint experimentation; Operational Test and Evaluation Force overview of responsibilities and perspectives on experimentation; U.S. Air Force (USAF) briefings on environments and tools supporting USAF experimentation and transitioning USAF requirements into acquisition through experimentation; Carrier Group Four/Carrier Strike Force perspective on experimentation and training of East Coast battle groups; Navy Network Warfare Command overview and briefings on its role in experimentation; and discussion with First Marine Expeditionary Force/First Marine Expeditionary Brigade on recent experiences during Operation Enduring Freedom in Kandahar, Afghanistan, and how experimentation might have played a role.
- *August 15-16, 2002, in Washington, D.C.* Marine Corps Combat Development Command overview and briefings on Expeditionary Force Development System, Transformation and Concepts, Joint Concept Development Experimentation, and the Marine Corps Warfighting Laboratory; MITRE Corporation briefings on recent assessments of naval experimentation; and Army Director of Information Operations, Networks, and Space and Director of Information Systems for Command, Control, Communications, and Computers insights into Army experimentation efforts.
- *September 5-6, 2002, in Washington, D.C.* Planning meeting for report chapter captains and briefing by U.S. Navy on alternative acquisition process.
- *September 9, 2002, in Washington, D.C.* Subcommittee meeting; briefings by Air Force Experimentation Office.
- *October 23-24, 2002, in San Diego, California.* Subcommittee site visits to the USS *Coronado*, then the flagship of the U.S. Third Fleet, and the Marine Corps Tactical Systems Support Activity.

• *September 16-20, 2002, in Woods Hole, Massachusetts.* Committee deliberations and report drafting.

The months between the last meeting and publication of the report were spent preparing the draft manuscript, gathering additional information, reviewing and responding to the external review comments, editing the report, and subjecting the report to security review.

Acknowledgment of Reviewers

This report has been reviewed in draft form by individuals chosen for their diverse perspectives and technical expertise, in accordance with procedures approved by the National Research Council's (NRC's) Report Review Committee. The purpose of this independent review is to provide candid and critical comments that will assist the institution in making its published report as sound as possible and to ensure that the report meets institutional standards for objectivity, evidence, and responsiveness to the study charge. The review comments and draft manuscript remain confidential to protect the integrity of the deliberative process. We wish to thank the following individuals for their review of this report:

Herb Browne, Armed Forces Communications and Electronics Association,
Roy C. Evans, MITRE Corporation,
Paul G. Kaminski, Technovations, Inc.,
Bruce B. Knutson, LtGen, U.S. Marine Corps (retired),
John E. Morrison, Institute for Defense Analyses, and
Janos Sztipanovits, Vanderbilt University.

Although the reviewers listed above provided many constructive comments and suggestions, they were not asked to endorse the conclusions or recommendations, nor did they see the final draft of the report before its release. The review of this report was overseen by Lee M. Hunt, Alexandria, Virginia. Appointed by the National Research Council, he was responsible for making certain that an independent examination of this report was carried out in accordance with institutional procedures and that all review comments were carefully considered. Responsibility for the final content of this report rests entirely with the authoring committee and the institution.

Contents

Executive Summary

The nation's military Services have long embraced experimentation as a fundamental tool for force development. For the U.S. Navy and the U.S. Marine Corps in particular, experimentation has enabled historical transformations in the fleet[1] such as naval aviation, for example, and more recently has contributed to a better understanding of and appreciation for emerging operational concepts such as urban warfare and network-centric operations.[2] Furthermore, in its *Naval Transformation Roadmap*,[3] the Department of the Navy has identified experimentation as a key enabler to achieve its vision of the future naval forces. Given the intended use of experimentation, the Department of the Navy must ask: Is naval experimentation as good as it needs to be? Is it as good as it can be? Answering these questions provides an underlying motivation for this study.[4]

---

[1]The term "fleet" is used in this report to include both the U.S. Navy's fleet and the operating forces of the U.S. Marine Corps.

[2]Because network-centric operations is a new defining concept that uses the information network rather than major platforms as an underlying framework for force structure and operations, experimentation should play a central role in transitioning naval forces to network-centric-based operations. See Naval Studies Board, National Research Council, 2000, *Network-Centric Naval Forces: A Transition Strategy for Enhancing Operational Capabilities,* National Academy Press, Washington, D.C.

[3]Secretary of the Navy Gordon England, Chief of Naval Operations Vern Clark, and Commandant of the Marine Corps James L. Jones. 2002. *Naval Transformation Roadmap: Power and Access . . . From the Sea,* Department of the Navy, Washington, D.C.

[4]In an earlier report, the Naval Studies Board expressed concern about the adequacy of the Navy and Marine Corps approach to experimentation, citing a tendency to focus on a few critical events, an

CONTEXT FOR THE REPORT

The Chief of Naval Operations (CNO) and the Commandant, U.S. Marine Corps (CMC) have defined their Service capstone concepts in "Sea Power 21"[5] and *Expeditionary Maneuver Warfare,*[6] respectively.

For the Navy, the CNO recently established an organizational policy for experimentation in "Sea Power 21" that is captured under the innovation element called Sea Trial, ". . . a continual process of rapid concept and technology development."[7] While Sea Trial is fleet-led under the guidance of the Commander, Fleet Forces Command (CFFC), this organizational policy places the Navy Warfare Development Command (NWDC) in a central role:

> The Commander, U.S. Fleet Forces Command, will serve as Executive Agent for Sea Trial, with Second and Third Fleet commanders sponsoring the development of Sea Strike, Sea Shield and Sea Basing capabilities. . . . The Systems Commands and Program Executive Offices will be integral partners in this effort. . . . The Navy Warfare Development Command, reporting directly to the Commander, U.S. Fleet Forces Command, will coordinate Sea Trial.[8]

For the Marine Corps, force development and requirements to be determined through experimentation remain the responsibility of the Marine Corps Combat Development Command (MCCDC). The MCCDC has now established the Expeditionary Force Development Center to continue to develop concepts, coordinate assessment and experimentation, and integrate the implementation of doctrine, organization, training, materiel, leadership, personnel, and facilities (DOTMLPF) across the range of Marine Corps operations.[9] The underlying purpose of establishing this center, however, is to develop a single process—the Expeditionary Force Development System—by which the Marine Corps will transform its force

---

extreme underutilization of analysis and of modeling and simulation, and a failure to decompose broad problems into components that can be studied in appropriate ways over time. See Naval Studies Board, National Research Council, 2000, *Network-Centric Naval Forces: A Transition Strategy for Enhancing Operational Capabilities,* National Academy Press, Washington, D.C.

[5]ADM Vern Clark, USN. 2002. "Sea Power 21: Projecting Decisive Joint Capabilities," *U.S. Naval Institute Proceedings*, Vol. 128, No. 10, October 1, pp. 32-41.

[6]Gen James L. Jones, USMC, Commandant of the Marine Corps. 2001. *Expeditionary Maneuver Warfare: Marine Corps Capstone Concept,* Warfighting Development Integration Division, Marine Corps Combat Development Command, Quantico, Va., November 10.

[7]Navy Warfare Development Command. 2003. "Sea Power 21," Newport, R.I. Available online at <http://www.nwdc.navy.mil/SeaPower21.asp>. Accessed November 9, 2003.

[8]ADM Vern Clark, USN. 2002. "Sea Power 21: Projecting Decisive Joint Capabilities," *U.S. Naval Institute Proceedings,* Vol. 128, No. 10, October 1, p. 39.

[9]Col Frank DiFalco, USMC, Joint Concept Development and Experimentation Operations Center, Marine Corps Combat Development Command, "Marine Corps Role in JCDE," presentation to the committee on August 15, 2002.

in order to be able to carry out Expeditionary Maneuver Warfare, its capstone concept.

Given the charters of the NWDC and the MCCDC, the committee viewed these organizations as central to its examination of naval experimentation. It focused on their processes, programs, and successes to date, as well as on organizations with which the two interact and that are important participants in experimentation, such as the Office of Naval Research, the Third Fleet, the Navy Network Warfare Command, and the U.S. Joint Forces Command (USJFCOM). The experimentation activities of these organizations collectively bring together technologies, systems, doctrines, and tactics, techniques, and procedures (TTPs) that cut across traditional boundaries and cultures, that require substantial integration efforts, and that have the potential to dramatically improve naval capabilities. The NWDC and the MCCDC also conduct experimentation that supports immediate and mid-term needs in the fleet, and in addition are responsible for coordinating their Services' participation in joint experimentation.

Given the number of organizations involved in naval experimentation and the range of resulting interpretations, approaches, responsibilities, and activities, the committee defined "experimentation" as follows: *Military experimentation is an activity conducted to discover, test, demonstrate, or explore future military concepts, organizations, and equipment and the interplay among them, using a combination of actual, simulated, and surrogate forces and equipment.*

The definition communicates two important points. First, experimentation explores more than equipment. It includes the development of doctrine, organization, training, materiel, leadership, personnel, and facilities that collectively constitute the mission capability of a military force. Second, experimentation encompasses a spectrum of activities, such as studies and analyses, seminars and conferences, work by subject-matter experts, war games, modeling and simulations, and experiments that are small and focused, as well as large field events with live forces. The need for this range of activities in conjunction with well-structured experimentation campaigns[10]—to investigate multiple concepts and alternative paths, explore fuzzy spaces, understand negative results, and discard and/or endorse solutions in order to "write the book" on how to achieve a transformational capability—is a basic tenet of this study that was used by the committee to assess current Navy and Marine Corps experimentation efforts.

---

[10]The committee defined an experimentation campaign as *a planned and cohesive, multiyear program of experimentation built on a series of experiments and related activities to develop the knowledge needed to inform major decisions about future forces, explore the viability of potential or planned changes to forces or their capabilities, and/or confirm that planned capability development and directions will enable forces to perform as expected.* See Chapter 2 in this report for additional details about the characteristics of campaigns, their value, and their application of spiral methodology.

CURRENT STATUS OF NAVY AND MARINE CORPS EXPERIMENTATION

The committee noted some successes resulting from recent naval experimentation, as managed both by the NWDC and the MCCDC. Under the NWDC, experimentation, including the large and complex Fleet Battle Experiments (FBEs) Alpha through India, has addressed important concepts and topics, such as network-centric operations for naval and joint fire power, theater and air missile defense, precision engagement, time-critical strike, and defense against asymmetric threats. The committee's summary observation is that recent Navy experimentation has indeed demonstrated the feasibility of some new operational concepts using surrogate or prototype or existing systems, and has led to the adoption by the fleet of new TTPs and the development of new doctrine for fleet operations.

The MCCDC, including the Marine Corps Warfighting Laboratory, can also point to successes. A review of results derived from experimentation beginning with the Hunter Warrior campaign in 1997 shows that concepts, doctrine, and TTPs resulting from experimentation have transitioned successfully to forces in the field and that experimentation has resulted in changes in minor equipment items that are in the field today. Here, too, experimentation was used to address important areas of investigation, such as nonlethal weapons, small unmanned aerial vehicles, and precision targeting, and to explore both desert and urban environments. A notable success coming from the Urban Warrior campaign of 1998 influenced major changes in doctrine for operations in urban terrain.

There is also progress in joint experimentation. Both the NWDC and the MCCDC are focal points for their respective Services' participation in joint experimentation. What the committee noted was more "thinking joint"—more collaboration on joint concept development, more experiment planning with a joint context in mind, and increasing participation in USJFCOM-sponsored joint experimentation and its attendant processes.

Given these positive results, the committee believes that both Navy and Marine Corps experimentation is enabling learning and producing meaningful results directed at promising concepts and technologies in a number of key naval mission areas. Nonetheless, these are modest successes, and the questions remain—Is naval experimentation as good as it needs to be for achieving the challenges of Sea Power 21 and Expeditionary Maneuver Warfare? Or is it as good as it can be? Here the committee would answer "no" to both questions. The Naval Services will have to enhance their programs and specifically, to improve their strategies, mechanisms, and processes for conducting experimentation. Alternately stated, with the moderate and careful changes that are the subject of the committee's recommendations, the effectiveness of experimentation for shaping future naval forces can be significantly improved.

RECOMMENDATIONS FOR IMPROVING THE OVERALL EFFECTIVENESS OF NAVAL EXPERIMENTATION

Establish Senior Navy Oversight and Annual Review of Experimentation Efforts

The CNO is using experimentation as an enabler for realizing the vision outlined in Sea Power 21, with the expectation that experimentation will contribute to building future forces, provide the means for development of advanced concepts, and facilitate the movement of capabilities to the field. What the committee sees is a modest program of experimentation managed by an organization (the NWDC) with insufficient influence over the Navy experimentation program[11] and its numerous participants; this insufficient influence extends to the funding and assets required and to the leverage needed for moving the results of experimentation either into acquisition or into programs of record through the requirements process. Even under the best of circumstances, when the results of experimentation provide ample evidence of the need for new capabilities, the bridges between the experimentation organization and the acquisition and requirements organizations of the Navy are fragile; they depend unduly on the exercise of coordinating skills and personal interactions.

The Navy's situation is not unique or unusual for a large bureaucratic institution. One lesson learned from past successes in experimentation—both by an earlier, historical Navy and by the other Services—suggests the need for strong oversight and participation by the most senior leadership. This level of involvement is necessary when experimentation is intended to result in significant new capabilities for the field.

The situation is not so exacerbated in the Marine Corps. The Marine Requirements Oversight Council (MROC) periodically reviews experimentation and effectively oversees its strategic direction and results. The ties to requirements and acquisition are less fragile, but they need strengthening. The smaller size of the Marine Corps facilitates tighter direction and control.

A mechanism is needed to ensure that experimentation is consistent with the CNO's vision and direction while garnering the support required Navy-wide. The committee debated but elected *not* to recommend a realignment of the NWDC under the CNO, given the NWDC's recent reorganization under the CFFC. Instead, the committee recommends an annual review of experimentation to engage the most senior leadership—a review that is not pro forma and that

[11]There is no single formal Navy experimentation program. Instead, a number of organizations are engaged in experimentation activities that collectively constitute a Navy program. In Chapter 3, see the subsection entitled "Organizational Roles and Major Participants in Navy Experimentation" for a discussion of various Navy organizations involved in experimentation.

provides an integrated view of experimentation activities across the Navy, including linkages to joint experimentation.

Recommendation 1: To ensure that experimentation is a key enabler of his long-term vision, the Chief of Naval Operations should establish and, together with the Vice Chief of Naval Operations, participate in an annual review of the experimentation program with the senior leadership of the Navy. This process should make visible the program content; the balance of near-term, mid-term, and long-term objectives; the progress that has occurred to date; the results that have been achieved; the use that has been made of the results relative to DOTMLPF, including the transition to requirements, acquisitions, programs of record, and/or capabilities in the fleet; and guidance for the future.

Strengthen Transition Processes

The mechanisms and processes for transitioning the results of experimentation, including transition planning, need strengthening. For the Navy, despite the transitioning of some concepts, doctrine, and TTPs to the field, there is little evidence that equipment and capabilities resulting from experimentation have transitioned or directly linked to major acquisitions and programs of record. Although the results of experimentation have had some influence on acquisition designs (e.g., for the Littoral Combat Ship), these instances appear to have evolved through personal connections rather than through institutional mechanisms. In general, institutional mechanisms are preferred because they typically endure beyond personal connections.[12]

In the Marine Corps, more structured transition planning is evident, as are some successes in moving concepts, doctrine, TTPs, and minor items of equipment into the field. Nonetheless, there was no evidence that major items of equipment from experiments had resulted in major acquisitions or displaced programs of record. Although not all experimental capabilities warrant transition in the context of cost, risk, and military value, the debate should be allowed. The committee also noted that spiral development—a potential enabler for transitioning capabilities more rapidly to the fleet—has not been explored systematically or incorporated as a fundamental method of experimentation.

Recommendation 2: To strengthen the transition of experimentation results to the requirements and acquisition processes, the Secretary of the Navy, the Chief of Naval Operations, and the Commandant of the Marine Corps should institute

[12]However, there are times when a personal connection is the initial enabler and the CNO can often assign or rotate key personnel to maximize benefits.

specific procedures to facilitate and accelerate the transition of capabilities identified through experimentation to the fleet.

Specifically, for the Navy:

- The Commander, Fleet Forces Command; Deputy Chief of Naval Operations for Warfare Requirements and Programs (N6/N7); Deputy Chief of Naval Operations for Resources, Requirements, and Assessments (N8); and Assistant Secretary of the Navy for Research, Development, and Acquisition should collectively formalize a planning process for the transition of the operational and system capabilities emerging from experiments to the fleet. The process should include framing transition issues and identifying potential funding gaps.
- The N6/N7 and the N8 should develop a process which ensures that the successful results of experiments are adequately evaluated and competed with the programs of record in the context of cost, risk, and military value.
- The Navy operational and acquisition communities should explore means to accelerate transition of the results of experimentation to the fleet more aggressively. These means should include the expanded use of other transaction authority and spiral development.[13]
- The Navy test community should explore new roles for the Operational Test and Evaluation Force, including its early participation in the experimentation program, with its advisory assessments provided directly to experiment managers.

Specifically, for the Marine Corps:

- The Marine Corps Combat Development Command, in conjunction with the Marine Corps Systems Command, should expand early transition planning in order to include the framing of transition issues and the identification of potential funding gaps.
- The Marine Requirements Oversight Council should establish a process which ensures that the successful results of experiments are adequately evaluated and competed with the programs of record in the context of cost, risk, and military value.
- The Marine Corps operational and acquisition communities should explore means to accelerate transition of the results of experimentation to the fleet more aggressively. These means should include expanded use of other transaction authority and spiral development.

---

[13]The use of spiral development to accelerate capabilities to the fleet has not been systematic to date, although spiral development is a component of "Sea Power 21." See Chapter 5 for additional

- The Marine Corps test community should explore new roles for the Marine Corps Operational Test and Evaluation Activity emphasizing its early participation in the experimentation program, with its advisory assessments provided directly to experiment managers.

Enhance the Naval Experimentation Programs

Certain important areas are not yet adequately explored in the naval experimentation programs, although some of these areas are gaining definition.[14] For the Navy, these omissions are due in part to its approach to experimentation, which in the past has not been founded on sufficiently robust experimentation campaigns but on an over-reliance on many individual events such as FBEs. Such singular events cannot provide the depth of knowledge required to explore potential concepts and capabilities sufficiently. Not only has the breadth of these programs been limited, but the number of concepts explored has been small, the concepts have not covered a sufficiently broad range, and they have not been systematically chosen and developed. As a result, the experimentation programs have lacked the cohesion and comprehensiveness needed to address the challenges of Sea Power 21 or to deal conclusively with questions about capabilities that will be delivered by the programs of record.

Areas that need further investigation include over-the-horizon, time-critical strike; use of extended-range guided munitions for long-distance, high-volume, rapid fire support; expanded applications of network-centric capability to deployable undersea sensor arrays; mine/countermine warfare; and the use of unmanned aerial vehicles to locate and identify targets. The Navy's experimental work to date has brought out overarching issues such as the need to achieve a satisfactory common operating picture; de-confliction; and bandwidth size and management. Some important areas not yet explored include Vertical Launch System reloading at sea, assault breaching of mine or obstacle fields near and on

---

details. Both the Air Force and Army have enjoyed some notable successes by incorporating it into their respective experimentation programs. See Chapter 3 for additional details. Given the Navy's emphasis on network-centric operations and NETWARCOM's emerging role, the committee believes that spiral development should be explored through experimentation to accelerate network-centric capabilities into operations. There are also naval infrastructures that may support such an exploration—namely, the Navy's Distributed Engineering Plant and the Marine Corps' Tactical Systems Support Activity.

[14]In January 2003, the CNO requested that the CFFC—as part of its lead role for Sea Trial in support of Sea Power 21—"[d]raft and implement a comprehensive roadmap (by May 2003) that integrates studies, wargames, experimentation, and exercises with evaluation metrics and an execution timeline." See Chief of Naval Operations, 2003, *CNO Guidance for 2003,* Department of the Navy, Washington, D.C., January 3. Available online at <http://www.chinfo.navy.mil/navpalib/cno/clark-guidance2003.html>. Accessed November 9, 2003.

the beach, and continued decisive operations under impaired network conditions and unfavorable environmental conditions.

In response to the CNO's guidance, the CFFC, through the NWDC, recently drafted the Sea Trial experimentation campaign plan.[15] The committee believes that this is a step in the right direction, although the impact of the plan on the Navy is as yet unclear.

For the Marine Corps, there has been a shift in recent years from a balanced program of experimentation campaigns to a program of experimentation based on near- and mid-term objectives. While these immediate challenges are important, there remains a need for continuing investment in long-term experimentation. Examples include sea basing, for which a program of experimentation needs to be designed, funded, and executed with the objective of realizing new capabilities, doctrine, and TTPs; operation in brown-water littorals to negate potential threats; and unconventional warfare, which will require the adaptation of current procedures for use from a sea base in brown-water operations.

Recommendation 3: To address strategic, long-term objectives of Sea Power 21 and Expeditionary Maneuver Warfare, the Department of the Navy should expand its programs for experimentation.

Specifically, for the Navy:

- The Commander, Fleet Forces Command, with the support of the Navy Warfare Development Command, should (1) create and maintain updated experimentation campaigns that address transformation objectives while identifying actionable steps and the organizations responsible for them; (2) ensure a balance in experimentation efforts directed at near-, mid-, and long-term objectives; (3) conduct experimentation sufficient to ensure that the highest-priority operational concepts are explored adequately for incorporation into the fleet and its operations; (4) establish adequate mechanisms for continued improvements and modifications to the experimentation program; and (5) maximize the effectiveness of joint experimentation in accordance with Recommendation 7 (below).
- The Navy Network Warfare Command and its supporting organizations should play a lead role in coordinating the information network aspects of experimentation and in enabling the realization of network-centric capabilities for the fleet through related concept development or exploration and spiral development processes.

[15]Commander, Fleet Forces Command. 2003. *Sea Trial—Concept Development and Experimentation Campaign Plan (U),* Working Paper (draft), Norfolk, Va., May (Classified).

Specifically, for the Marine Corps:

- In collaboration with the Navy Warfare Development Command, the Marine Corps Combat Development Command should augment its experimentation program by developing experimentation campaigns to address its strategic, long-term objectives, such as sea basing, conventional and unconventional expeditionary warfare, and, jointly with the Navy, assured access. Furthermore, its overall experimentation campaign should encompass all levels of force structure and activity necessary to meet the range of potential threats and future operational demands.

As requested in the terms of reference for this study, a set of specific enhancements recommended for the naval experimentation programs is provided.[16] These enhancements address future challenges stemming from Sea Power 21 and Expeditionary Maneuver Warfare and elaborate on areas for experimentation toward future capabilities in the naval programs of record.

Enhance Navy Experimentation Processes

Unlike the Marine Corps experimentation program, the Navy experimentation program needs more robust methods to build experimentation campaigns. To date, preparing for the event of an FBE has largely (but not exclusively) been a focus of activities. The NWDC needs both to shift focus and to augment processes.

The NWDC requires enhanced processes for the following purposes: to select concepts for exploration; to integrate more overarching studies and analyses throughout the events of its campaigns; to build, mature, and evaluate concepts; and to maximize and apply a full range of experimentation venues (such as games, modeling and simulation, and limited objective experiments) in a more systematic manner. The NWDC's experimentation campaigns require expansion and structure in order to ensure thorough learning of the relevant phenomena and of capability options and extensions of these by studies, analyses, and "writing the book" on concepts and capabilities under investigation so as to build definitive knowledge. Large field experiments are important and provide value, but they should be used only when appropriate—for instance, in exploring scalability or integration issues, or when warfighters need to interact with potential new capabilities. When experiments are introduced into large-scale exercises, conflicts between operational readiness objectives and an experiment's objectives can arise, and the priorities of the exercises can degrade the value of the experiment. The recommended enhancements, as institutional processes, should pro-

[16]In Chapter 6, see the section entitled "Specific Enhancements for the Naval Programs of Experimentation."

duce a better balance across all venues and yield knowledge that cannot be obtained in large field experiments, such as that gained from a large-scale examination of options or excursions. These are necessary changes in strategy for the Sea Trial process. As noted above, the recent emergence of the Sea Trial Experimentation Campaign Plan is a step forward in this direction.

Recommendation 4: To improve the effectiveness of its experimentation efforts, the Navy Warfare Development Command should augment its end-to-end experimentation processes by making the following key changes:

- Expand the emphasis on experimentation campaigns that use a full spectrum of experimentation activities, with analysis integrated throughout the campaigns as well as applied to determine which venues are most appropriate.
- Conduct significantly greater amounts of systematic and innovative analysis earlier and throughout the experimentation process in order to select, develop, and broaden understanding of the operational concepts to be explored, including a range of multiple and competing concepts.
- Broaden the incremental, sequential approach by using spiral methodology. Apply the sequential approach to war games, to modeling and simulation, and to small-scale, more narrowly focused experiments. Build larger-scale experiments on the basis of the results of such a sequential approach.
- Establish a standing, high-level, independent technical advisory board composed of experts in methods of innovation and experimentation and reporting to the Commander of the Navy Warfare Development Command, as a means to foster more robust experimental processes, maintain their quality, and make recommendations for improvements.

Sustain and Use Navy Experimentation Resources More Effectively

The committee questioned whether sufficient resources are available to the NWDC for experimentation but found this difficult to answer for two reasons. One involves the past emphasis on large fleet events, compounded more recently by joint experimentation. Large fleet events are costly, require many months of effort, involve competing needs for resources, and can be less effective at answering important questions than other venues to experimentation are. However, in the committee's judgment, maximizing the use of smaller and more effective venues (recommended above) provides better use of the resources available and represents a sound change in methodology as well. The committee applauds the recent Sea Trial Experimentation Campaign Plan for its inclusion of a greater number of small experimentation venues.

The second source of difficulty in answering the funding question is that the NWDC lacks line-item funding to cover the full costs of its own experimentation program. Rather, it coordinates and leverages both participation and funding from many organizations. In addition to its own funding, the NWDC relies on supplements from various organizations such as the Office of Naval Research (ONR). Although ONR has intended to provide some $20 million to $40 million annually for experimentation, in the past it has reallocated funds to other, higher-priority efforts. Such diversion of funds has occurred with other sponsors as well. These perturbations severely impact planning, preparation, conduct, and evaluation of experiments such as FBEs, which typically have cycle times of 12 to 18 months. When funds are reallocated, undue effort is expended by the NWDC on finding alternative funding, to the detriment of planning and conducting experimentation.

What is clear is that the Navy is not making the most effective use of the experimentation resources already committed. The Navy should establish a line item for funding that reflects the value it places on experimentation, stands on its own merits, and is sustained and supported Navy-wide. Moreover FBEs will continue to be an important component of experimentation when used more appropriately; however, resource contention, such as from sharing assets with training exercises, needs to be addressed. The CFFC's command of fleet assets should provide for resolution and leverage in this matter.

Recommendation 5: The Navy should use the resources already programmed for experimentation more effectively, while also resolving resource contention surrounding experimentation. Specifically:

- The Chief of Naval Operations, with input from the Vice Chief of Naval Operations; the Commander, Fleet Forces Command (CFFC); the Deputy Chief of Naval Operations for Warfare Requirements and Programs (N6/N7); and the Deputy Chief of Naval Operations for Resources, Requirements, and Assessments (N8), should determine, and then plan, program, and preserve sufficient funding for the experimentation program within the Future Year Defense Program as a matter of priority. Funding for the Navy's experimentation program should be a separate line item that is not commingled with the funds provided by other Navy sponsors, such as the Office of Naval Research.
- The CFFC should ensure that sufficient priority is given to experimentation needs when fleet training exercises and maintenance events override or threaten to compromise experiments. The CFFC should oversee and adjudicate conflicts that arise during all stages of fleet experiments.
- In collaboration with the CFFC and the Navy Warfare Development Command (NWDC), the Chief of Naval Education and Training, the Naval Postgraduate School at Monterey, California, and the Naval War College at Newport, Rhode Island, should develop a tailored course on experimentation in order to train and educate sailors, Marines, and civilians, and to instantiate best practices in experimentation.

Enhance Infrastructure and Tools for Naval Experimentation

The Naval Services are in need of improvements in the infrastructure and tools supporting experimentation. Some critical platforms for experimentation are unavailable, including ship platforms—a shortfall compounded by the potential decommissioning of the USS *Coronado*—and airborne command, control, communications, computers, intelligence, surveillance, and reconnaissance (C4ISR) assets, some of which are currently under the ownership or control of the other Services.

In addition, the Naval Services will require improvements in modeling and simulation (M&S) tools to explore the full range of experimentation venues, including tools for generating scenarios, populating databases, and collecting and analyzing data. Simulations are required to explore more tactical-level interactions, and, at the conclusion of the Millennium Challenge '02 exercise, USJFCOM noted that the Defense Department's existing M&S capabilities are insufficient to represent future operational concepts. In response, USJFCOM is proceeding with the expansion of its own Continuous Experimentation Environment. As a result, the Navy and Marine Corps will have an opportunity to leverage the USJFCOM investments while also ensuring compatibility across the respective Service and joint environments. Industry partnerships can be explored also, as appropriate.

Recommendation 6: To investigate a full range of experimentation venues, the Department of the Navy should enhance its infrastructure and tools for naval experimentation.

Specifically for the Department of the Navy:

- Given the importance of having a command ship platform with an expert and experienced cadre of experimentalists for work in network-centric operations, the Department of the Navy should ensure that the USS *Coronado* (or a comparable platform) remains available for experimentation as it makes changes in the fleet of ships in commission or in active reserve duty.
- The Department of the Navy should augment its modeling and simulation tools and infrastructure to support a full spectrum of naval and joint experimentation campaign activities (e.g., concept development, war games, and limited-objective experiments) to support tactical-level interactions and to reflect next-generation warfighting environments. It should also supplement its tools for building, validating, and verifying models; for generating scenarios and populating databases; and for collecting and analyzing data, while ensuring tools that can function and integrate within the various frameworks and environments as future experimentation campaigns are defined and executed. In addressing enhancement of its infrastructure and tools for naval experimentation, the Department of

the Navy should leverage the capabilities of other organizations, such as the U.S. Joint Forces Command.

Balance Naval and Joint Experimentation

As joint operations continue to leverage naval capabilities, planning is needed to link naval and joint experimentation across a spectrum of activities ranging from the earliest concept development through analyses, war games, and simulations, leading ultimately to limited-objective experiments and larger fleet experiments. Given this requirement, the committee views the current state in naval experimentation as having both limitations and opportunities.

Joint concept development requires more synergistic collaboration with the Naval Services than is now taking place. Concept development for the Navy and the Marine Corps is focused at the NWDC and the MCCDC, respectively, and on the joint side at the USJFCOM. Although these organizations are interacting, the committee believes that the interaction is not as close or extensive as it needs to be. While the development of joint concepts does involve the Naval Services, it does not appear to build on the Services' concepts in a substantive way; nor is there a good, detailed crosswalk between the joint concepts and those developed by the Naval Services.

In the past the Navy and Marine Corps have participated with USJFCOM in a number of joint experimentation events ranging from war games to field experiments. Attention has been given more recently to the large-scale field experiment Millennium Challenge '02, and USJFCOM is currently planning a significant program of future experimentation activities. The Naval Services are expected to be active participants and should play a substantive role in defining these events.

There are opportunities to expand joint experimentation with the Combatant Commands. Both the Navy and the Marine Corps have participated in conducting experiments with the Combatant Commands, usually in conjunction with exercises such as Kernel Blitz. Past experiences have been valuable, particularly with respect to advances in command and control, communications, developing and refining procedures and working with prototype systems, and the involvement of coalition nations. There is a need for a coordination mechanism that will operate between the Naval Services and the Combatant Commands to identify and build programs of mutual interest. One alternative is to use this coordination mechanism between the Combatant Command and the Service component commands assigned to it.[17] The component commands could in turn interact with appropriate elements of their respective Services (e.g., with the NWDC and the MCCDC).

[17]Such as either the appropriate Navy fleet command or the numbered fleet commands under that fleet command.

More cross-Service experimentation is required, yet interactions among Service concept development centers appear to be limited. The Navy has involved both the Army and the Air Force in past FBEs, and the NWDC previously indicated a desire to increase other Services' involvement in future FBEs. The current Marine Corps experiments do not appear to have a significant joint or cross-Service perspective, yet cross-Service experimentation involving two or more Services is ideally suited for exploring operations at the tactical level. Moreover, there is a pressing need to investigate joint interactions at the tactical level, given the growing intensity of such interactions in recent operations. Expanding cross-Service experimentation, however, requires a mechanism for coordinating force deployment schedules—although it is conceivable that such coordination could be carried out at the top levels of the Services, with USJFCOM serving as an intermediary.

Recommendation 7: To ensure better preparation for future joint operations and to maximize the effectiveness of its participation in joint experimentation, the Department of the Navy should establish a set of principles and guidelines to balance experimentation requirements for Service-unique and joint experimentation, and it should then align and synchronize its participation accordingly. Specifically:

- The Chief of Naval Operations and the Commandant of the Marine Corps should actively support joint experimentation on the basis of a clear understanding of the priorities of the various joint concept development and experimentation activities. In addition, they should advocate top-level interest in operational concepts driven by naval force capabilities as well as concepts suitable in other operational environments.
- The Commander, Fleet Forces Command (CFFC) and the Marine Corps Combat Development Command (MCCDC) should conduct enough naval experimentation campaigns to ensure that the highest-priority naval operational concepts are adequately explored.
- The CFFC and the MCCDC should design all naval experiments with full recognition that the Navy and the Marine Corps will most likely be operating in a joint context; to the maximum degree feasible, the Naval Services should partner with the other Services in experimenting with relevant assets.
- The NWDC and the Marine Corps Warfighting Laboratory should achieve adequate cross-fertilization of joint and naval-specific operational concepts through substantive interaction of the respective concept development communities.
- The CFFC and the MCCDC should support and participate in joint experiments to explore the interaction and mutual support of the future operational concepts of each of the Services. These efforts should include staff interactions at the operational level and the "removal of seams" between components at the tactical level.

- The CFFC and the MCCDC should work with USJFCOM to identify key challenges (e.g., cruise missile defense, joint intelligence, surveillance, and reconnaissance) on which they could welcome joint and/or other Service contributions.
- The CFFC, N7, N8, and the MCCDC should examine the tradeoffs in the benefits of large, resource-intensive and less well controlled experiments against the opportunities lost by not conducting a greater number of smaller, more focused joint and/or naval experimentation activities.
- The CFFC and the MCCDC should systematically develop programs of joint experimentation with the Combatant Commands and should establish a coordination mechanism to facilitate the development of such programs.
- The CFFC and the MCCDC should increase cross-Service experimentation, particularly at the tactical level, and should establish a scheduling mechanism to facilitate this experimentation.

These principles are supplemented with a mission-based strategy for determining where to focus naval and joint experimentation. The mission-based approach is elaborated in Chapter 6, in the subsection titled "A Mission-Based Approach for Balancing Naval and Joint Experimentation."

1

Introduction

THE IMPORTANCE OF EXPERIMENTATION

Experimentation is a fundamental part of the scientific method and of the processes that move science into technology and utilization by modern society. Experimentation with large or small military units has always been a key tool in the shaping of military forces. Many military commanders have taken innovative steps—some successful and some disastrous—under the pressure of the exigencies of battle. Experimentation in peacetime, without the pressures of battle and the potential consequences of winning or losing wars, offers military commanders the opportunity to test and to explore the value of new military systems or new ways of using existing or planned systems.

Through experimentation, military planners can learn how systems will perform under field conditions against the actual (e.g., captured) or simulated equipment of potential adversaries, and they can learn the shortcomings of our own military systems and ways of improving them. They can explore how various operating techniques will work against surrogate opponents who use operational methods and tactics different from our own. By simulating future systems, they can also learn how those systems will work in simulated combat environments and how to use forces equipped with such proposed systems. By such means they can explore new ideas and concepts for the use of variously composed and equipped forces against diverse anticipated threats, and they can learn how to integrate such forces on a large scale in the joint and combined force environment.

Innovative design and use of military systems and techniques in warfare have often led to revolutionary changes in how military forces are constituted and how they fight. Examples from the 20th century include the use of armored forces in land warfare, the use of submarines and sea-based aviation in naval warfare, the application of ballistic missiles to intercontinental attack, the use of nuclear submarines to form long-enduring and essentially invulnerable undersea tactical and strategic strike forces, and the use of space systems for observation, communication, and navigation. Peacetime experimentation under simulated battle conditions on a large or small scale can be treated as a rehearsal for warfare if such innovations are allowed and encouraged. The experimentation process has therefore become a key tool for transforming U.S. military forces and systems from those oriented to the Cold War to those capable of waging future war against a different and evolving array of threats to U.S. national security. For the Navy and the Marine Corps in particular, experimentation has enabled historical transformation in the fleet,[1] such as that brought about by naval aviation, and, more recently, has contributed to a better understanding and appreciation of emerging operational concepts such as urban warfare and network-centric operations.

THE NETWORK-CENTRIC WARFARE CHALLENGE

A new defining concept for naval—and indeed, joint force—warfare, driven by the information revolution of the last quarter of the 20th century, has been that of network-centric warfare,[2] the idea that the information network, rather than platforms (e.g., ships, and aircraft), provides the underlying framework for force structure and utilization. A recent report of the Naval Studies Board on network-centric warfare indicated the importance of experimentation to developing naval forces according to that new defining concept. It stated:

> Experimentation provides a means to explore alternative doctrine, operational concepts, and tactics that are enabled by new technologies or required by new situations. That is, new technologies or situations may call for different ways of conducting operations. But without actual operational experience in using the new technologies or in using existing technologies in new situations, experiments are the next best thing. . . .
>
> Although they can fail in their ability to find the right solution, experiments should always provide knowledge about the ramifications of new ideas and

---

[1]The term "fleet" is used in this report to include both the U.S. Navy's fleet and the U.S. Marine Corps's Operating Forces.

[2]For additional reading, see VADM Arthur K. Cebrowski , USN, and John J. Garstka, 1998, "Network-Centric Warfare: Its Origin and Future," *U.S. Naval Institute Proceedings*, Vol. 124, No. 1, January, pp. 28-35; David S. Alberts, John J. Gartska, and Frederick P. Stein, 1999, *Network Centric Warfare: Developing and Leveraging Information Superiority*, 2nd Edition (Revised), Department of Defense C4ISR Cooperative Research Program, Office of the Assistant Secretary of Defense (Networks and Information Integration), Washington, D.C.

technologies, to assist those who write requirements by reducing the likelihood that they will specify requirements for too much (something that cannot be achieved within reasonable bounds) or too little (improvement insufficient to justify development).[3]

The report also states:

> Essential as they are, analytical methods alone are insufficient for the design of systems of this [network-centric operations] complexity. Actual experimentation by the fleet and Marine force elements is required, to learn how legacy subsystems and their components will operate together with existing or testbed versions of new subsystems and components and to devise concepts of operation using the new and the legacy subsystems and components in the actual operational environment. When such a development process, part of what has been called spiral development, is used, new equipment and concepts can be incorporated into the fleet and the Marine forces based on validated concepts of operation.[4]

The report then recommends:

> The spiral development approach involving design-test-design of new software and equipment and model-test-model to devise new joint concepts and their testing in fleet and Marine units should be adopted as a standard mechanism for achieving network-centric operations systems.[5]

While the report provided recommendations on how experimentation could contribute to building network-centric capabilities, it also expressed serious concerns about the adequacy of the Navy and Marine Corps approach to experimentation, citing a tendency to focus on a few critical events; an extreme underutilization of analysis, modeling, and simulation; and a failure to decompose broad problems into components that can be studied in appropriate ways over time.[6] These same concerns are relayed through specific questions in the terms of reference for this study (see the preface) and are addressed in the responses of the Committee for the Role of Experimentation in Building Future Naval Forces.

---

[3]Naval Studies Board, National Research Council. 2000. *Network-Centric Naval Forces: A Transition Strategy for Enhancing Operational Capabilities,* National Academy Press, Washington, D.C., p. 294.

[4]Naval Studies Board, National Research Council. 2000. *Network-Centric Naval Forces: A Transition Strategy for Enhancing Operational Capabilities,* National Academy Press, Washington, D.C., p. 22.

[5]Naval Studies Board, National Research Council. 2000. *Network-Centric Naval Forces: A Transition Strategy for Enhancing Operational Capabilities,* National Academy Press, Washington, D.C., p. 23.

[6]Naval Studies Board, National Research Council. 2000. *Network-Centric Naval Forces: A Transition Strategy for Enhancing Operational Capabilities,* National Academy Press, Washington, D.C., Sections 2.5.5 and 2.6.

PAST USE OF, AND NEW SIGNIFICANCE FOR, NAVAL FORCE EXPERIMENTATION

In moving toward network-centric operations, the naval forces (and the other Services), are in fact looking toward experimentation to explore the new concepts of operation demanded by the developing post–Cold War threat and made possible by advancing information and related technologies.

Experimentation is not a new technique to the Navy and Marine Corps. As described later in this report, experimentation with launching and recovering aircraft from ships began only a few years after aircraft were invented. Although the technology then did not allow the aircraft to be major strike systems, these aircraft were initially useful for locating enemy forces and later became attack systems as the technology advanced to enable such operations. Similarly, although Marines had been landing from ships in small boats for many decades before World War II, the pressures of warfare led to the rapid development of prototypical modern amphibious landing systems to enable and support amphibious, over-the-shore operations. After the war, experimentation led to the modern fleet of nuclear-powered attack and strategic missile submarines: that experimentation involved nuclear reactors in submarines—first in simulated runs across the Atlantic using a reactor in the laboratory, and then on actual extended voyages using reactors in submarines. These investigations were followed by adoption of previously developed streamlined hull forms and advanced control systems to "fly" the submarines under water. The streamlined hull forms and control systems were developed before, and independently of, the nuclear work described here. The hull shape development also involved experiments earlier than the ones done by the nuclear submarines.

Beyond these few broad examples, experimentation has been a key tool supporting advances in all forms of naval warfare; advances in gunnery, guided missiles, and naval ship propulsion; and progress in all other activities related to the shaping and operation of naval forces. A part of such force development has been the evolutionary improvement of particular equipment, platforms, and major combat and support systems as technological advances and budgetary constraints have permitted. These advances have been somewhat characteristic of the spiral development approach that has become increasingly attractive in shaping today's forces. A key difference between the block improvements in systems such as ships, aircraft, and air defense missiles and the movement toward more rapid system evolution through what is now called spiral development[7] has been the more rapid advances in the underlying technologies that characterize the key systems that are at the heart of the new network-centric warfare paradigm.

[7]Edward C. Adridge, Under Secretary of Defense. 2002. "Evolutionary Acquisition and Spiral Development," Memorandum, Department of Defense, Washington, D.C., April 12.

Whereas the technology of aircraft engines and missile guidance, for example, may have taken years to change significantly, the technology of information gathering, manipulation, and communication changes within months. Naval forces thus must develop new and more responsive approaches to change in order not to be left behind by adhering to an older pace and outdated methods. In this context the concept of spiral development supported by continual experimentation has taken on new meaning and significance for future naval forces.

BROAD RANGE OF ACTIVITIES COVERED BY EXPERIMENTATION

As would be expected from an activity that has assumed increasing importance in naval force planning and development, the term "experimentation" tends to be used rather loosely to cover a broad range of activities. Additionally, different communities within the naval forces, such as the R&D community, the acquisition community, and the fleet commands, and even other Service forces, tend to think about experimentation according to the dictates of their own orientation and activities.

This gamut of activity has created various definitional inconsistencies in the field of developing and improving the current and future utility of naval (and other) forces and have even entered the legal requirements for creating and fielding the forces. These complexities of definition are discussed in detail in Chapter 2, where the committee's definition of experimentation is given, in context.

However the definitions are applied, the current focus on experimentation has the purpose of solving the force design problems raised by the changing environment for the building and use of military forces. Given their cost, experiments—especially those involving the use of forces in the field—must be carefully focused and designed to make the best use of the resources committed. In today's budget environment of competing priorities and with the limited time available for fielding effective new systems and forces under the pressure of terrorist and other threats to U.S. and allied security, concepts for systems and force design and operation cannot be adopted at random simply to see "what if"—instead, they must be developed deliberately to solve a pressing military problem.

Thus, the post–Cold War Navy strategy articulated in 1992 in *From the Sea*[8] was driven by the need to orient U.S. naval power toward the littorals, whence the new threats originate. The concept articulated in *Operational Maneuver From the Sea*[9] is meant to minimize the need for opposed landings over the beach and

[8]Department of the Navy. 1992. ". . . From the Sea," U.S. Government Printing Office, Washington, D.C., September.

[9]Headquarters, U.S. Marine Corps. 1996. *Operational Maneuver From the Sea,* U.S. Government Printing Office, Washington, D.C., January 4.

instead to have strikes from the sea launched against enemy inland "centers of gravity" with speed and surprise, as well as to hide the source of attack either by maneuvering far from these centers or deep within the sea or both. Advancing technologies for vertical-lift aircraft, higher-speed landing craft, and long-range fire support make this possible. Also emerging is the need for the sea basing of forces in order to avoid the risk of using land bases in an environment of shifting coalitions, when the stability and utility of any single base are uncertain and when fixed bases on land are vulnerable to attack from threats that range from terrorists to ballistic missiles. The Marine Corps has emphasized the importance of warfare in built-up areas considering these new kinds of threats, and the need to subdue opponents quickly in such areas to minimize both military and civilian casualties. The Marine Corps's Urban Warrior experimentation campaign provided useful lessons that could be applied in Operation Enduring Freedom.

POLICY FOR TRANSFORMING NAVAL FORCES THROUGH EXPERIMENTATION

From the Sea was just a first step in reorienting U.S. naval power. The need to project naval power at sea and inland on a global scale has led to a more complete change in naval force strategy articulated by the Chief of Naval Operations (CNO) in a 2002 document entitled "Sea Power 21: Projecting Decisive Joint Capabilities."[10] The new naval force strategy includes Sea Strike (the means by which Navy firepower and Marine ground forces will be projected where they are needed in order to protect and support U.S. and allied interests wherever they are threatened); Sea Shield (the means of protecting the naval forces at sea and ashore and in regions under threat to which they have been sent for support); Sea Basing (the means to support forces mainly from the sea without having to rely as much as in the past on land bases that may be considered politically intrusive and that would certainly be more vulnerable to attack by hostile forces); and FORCEnet, which integrates Sea Strike, Sea Shield, and Sea Basing as the "operational construct and architectural framework for naval warfare in the information age, integrating warriors, sensors, command and control, platforms, and weapons into a networked, distributed combat force."[11]

This evolution of naval force strategy has created a host of issues related to integrating new technologies and operational capabilities into a new kind of naval force system. The CNO has established the organizational policy for approaching

---

[10]ADM Vern Clark, USN. 2002. "Sea Power 21: Projecting Decisive Joint Capabilities," *U.S. Naval Institute Proceedings,* Vol. 128, No. 10, October 1.

[11]ADM Vern Clark, USN. 2002. "Sea Power 21: Projecting Decisive Joint Capabilities," *U.S. Naval Institute Proceedings,* Vol. 128, No. 10, October 1, p. 37.

these issues through Sea Trial, "a continual process of rapid concept and technology development."[12] The policy states:

> The Commander, U.S. Fleet Forces Command, will serve as Executive Agent for Sea Trial, with Second and Third Fleet commanders sponsoring the development of Sea Strike, Sea Shield, and Sea Basing capabilities. . . . The Systems Commands and Program Executive Offices will be integral partners in this effort. . . . The Navy Warfare Development Command, reporting directly to the Commander, U.S. Fleet Forces Command, will coordinate Sea Trial. Working closely with the fleets, technology development centers, and academic resources, the Navy Warfare Development Command will integrate wargaming, experimentation, and exercises to speed development of new concepts and technologies. They will do this by identifying candidates with the greatest potential to provide dramatic increases in warfighting capability. Embracing spiral development, these technologies and concepts will then be matured through targeted investment and guided through a process of rapid prototyping and fleet experimentation.[13]

The *Naval Transformation Roadmap* of the Department of the Navy further states:

> . . . Navy headquarters (OPNAV) will support Sea Strike, Sea Shield, and Sea Basing concept development by working directly with NWDC [Navy Warfare Development Command] and fleet elements to ensure operational priorities and lessons learned are accurately reflected in budgetary resourcing. Specifically, Mission Capabilities Package (MCP) teams under Deputy Chief of Naval Operations (Warfare Requirements and Programs) will assist in developing, resourcing, and implementation of these concepts, and in making the linkages from concepts and technologies to acquisition programs and fleet forces. [14]

Carrying this process forward, the Deputy Chief of Naval Operations (Warfare Requirements and Programs) has aligned Sea Strike, Sea Shield, Sea Basing, and FORCEnet with the Systems Commands by assigning implementation responsibility for these major elements of Sea Power 21 to the Naval Air Systems Command, the Naval Sea Systems Command, and the Space and Naval Warfare Systems Command.[15]

---

[12]Navy Warfare Development Command. 2003. "Sea Power 21," Newport, R.I. Available online at <http://www.nwdc.navy.mil/SeaPower21.asp>. Accessed November 9, 2003.

[13]ADM Vern Clark, USN. 2002. "Sea Power 21: Projecting Decisive Joint Capabilities," *U.S. Naval Institute Proceedings,* Vol. 128, No. 10, October 1, p. 39.

[14]Secretary of the Navy Gordon England, Chief of Naval Operations Vern Clark, and Commandant of the Marine Corps James L. Jones. 2002. *Naval Transformation Roadmap: Power and Access . . . From the Sea,* Department of the Navy, Washington, D.C., p. 34

[15]VADM John B. Nathman, USN, Deputy Chief of Naval Operations for Warfare Requirements and Programs, N7, "N6/N7 Naval Capabilities Development," presentation to the Naval Studies Board, November 6, 2002.

For the Marine Corps, force development and requirements to be determined through experimentation remain the responsibility of the Marine Corps Combat Development Command (MCCDC). More recently, MCCDC has established the Expeditionary Force Development Center to develop concepts, coordinate assessment and experimentation, and integrate the implementation of doctrine, organization, training, materiel, leadership, personnel, and facilities (DOTMLPF) across the range of Marine Corps operations.[16] The underlying purpose for this establishment, however, is to develop a single process—the Expeditionary Force Development System—by which the Marine Corps will transform itself in conformity with its capstone concept, Expeditionary Maneuver Warfare.

It is clear from the background presented above that as Navy and Marine Corps force transformation policy and practice have been developing, the issues and problems posed by the emerging U.S. naval force strategy will be approached to a major extent through experimentation with actual forces and military command structures. This consideration helped to determine the orientation and scope of this report.

SCOPE AND ORGANIZATION OF THIS REPORT

As stipulated in the terms of reference for this study, the committee concerned itself with the role of experimentation in building future naval forces in the joint environment. The study effort focused on programs of experimentation that explore warfighting concepts intended to meet the new threats and conditions of warfare facing U.S. naval forces. Given its objectives, the committee viewed NWDC and MCCDC as organizations central to its examination of naval experimentation and focused on their processes and programs and successes to date. The committee also focused on organizations with which NWDC and MCCDC interact and which have important roles in experimentation, such as the Office of Naval Research (ONR), the Third Fleet, the Navy Network Warfare Command (NETWARCOM), and the U.S. Joint Forces Command (USJFCOM). Their experimentation activities collectively bring together technologies, systems, doctrines, and tactics, techniques, and procedures that cut across traditional boundaries and cultures, that require substantial integration efforts, and that have the potential for dramatic improvements in naval capabilities. NWDC and MCCDC also conduct experimentation that supports immediate and mid-term needs in the fleet as well as forces in the fields, and they are also responsible for coordinating their Services' participation in joint experimentation. All of these experimenta-

[16]Col Frank DiFalco, USMC, Joint Concept Development and Experimentation Operations Center, Marine Corps Combat Development Command, "Marine Corps Role in JCDE," presentation to the committee on August 15, 2002.

tion activities—including the Navy's Fleet Battle Experiments Alpha through India (FBE-A through FBE-I), and the Marine Corps's experimentation efforts starting with the Hunter Warrior Campaign and continuing to the present time—were the focus of the committee's attention. It is recognized that the results of this study may be relevant not only for the Navy and Marine Corps but also for other naval organizations and their experimental efforts.

Within these bounds arose a number of issues and issue areas that are explored in this study. They include the following:

- The assessment of the effectiveness and utility of ongoing and future Navy and Marine Corps experimentation;
- Approaches to and problems of defining and planning an experiment campaign for a specific purpose;
- Planning for the incorporation of the results of successful experiments into the forces, especially when those results will affect the program of record;
- The availability of resources for experimentation and the interaction of experiments with exercises intended for other purposes when such exercises are used to provide ships and other resources needed for the experiments; and
- Developing an understanding of the relationship and balance between naval and joint experimentation.

Other issues are also highlighted and are discussed in context throughout the report as they arise.

Chapter 2, "Experimentation—What It Means," discusses issues in defining an experiment according to the needs and interests of different communities involved in naval force planning. It describes how individual experiments fit into a campaign of experimentation to explore naval force systems and concepts of operation, and it discusses methodological approaches to carefully designing experiments to yield the needed information. Chapter 3, "Experimentation—Past, Present, and Future," describes Navy, Marine Corps, and other Service experimentation and the results that have flowed from such efforts. Chapter 4, "Emerging Roles in Experimentation—The Joint Connection," discusses the movement toward joint experimentation, the benefits derived, the problems involved in conducting joint experiments using naval as well as other Services' forces, and the effects of joint experimentation on the Navy's and the Marine Corps's ability to conduct experiment programs to reinforce their own core military competencies. The results of Navy and Marine Corps experimentation programs and their bearing on the issues sketched above, as well as other issues that have emerged from this review, are evaluated in Chapter 5, "Effectiveness of Experimentation for Future Naval Capabilities." Chapter 6, "Recommendations for Improving the Overall Effectiveness of Naval Experimentation," presents the committee's recommendations.

ADDRESSING THE TERMS OF REFERENCE

The terms of reference for this study chartered the committee to examine the role of experimentation in enabling future naval forces to operate in the joint environment, and to address the opportunities offered by and the implications of experimentation. Within these broad areas, specific questions were also provided as part of the committee's charge. (The questions are listed in the section entitled "Terms of Reference" in the preface.)

Pointers to key sections of the report that address each aspect of the committee's charge are provided here for the reader's convenience.

- *Review of opportunities offered by experimentation.* Chapter 1 articulates a broad motivation for military experimentation and, more specifically, experimentation to support naval transformation. Chapter 3 provides two Navy case studies, one historical and one more recent, that illustrate the influence of experimentation on today's naval operations. The chapter also gives anecdotal historical examples that highlight the role of experimentation in evolving the capabilities of each of the other Services.
- *Review of implications of experimentation.* Chapter 2 discusses what it means to experiment in the context of military operations. It distinguishes among the various events that constitute a spectrum of experimentation activities, of which field experiments are one component. The chapter examines necessary processes for experimentation, including that of spiral development and the use of experimentation campaigns. It also delineates an environment and tools that support a sound experimentation program.
- *Review of what has been learned from experimentation and which results have transitioned.* Chapter 3 discusses what has been learned through recent experimentation programs of the NWDC and MCCDC beginning with the Navy's Fleet Battle Experiments Alpha through India, and the Marine Corps's Hunter Warrior, Urban Warrior, and Capable Warrior Campaigns. Chapter 3 also summarizes which results have transitioned, whether through concept, doctrine, and tactics, techniques, and procedures (TTPs), and/or acquisitions for the fleet and the field. In addition, Chapter 4 reviews joint experimentation and summarizes what has transitioned to date from this process. Chapter 5 assesses the adequacy of what has been learned and of the transition processes for naval experimentation, referring to and summarizing details from Chapter 3.
- *Review of spiral development.* Spiral methodology, including spiral development, is defined and discussed in the context of military experimentation in Chapter 2. Navy and Marine Corps use of spiral development in recent programs of experimentation is assessed in Chapter 5. Since its application by naval experimenters has not been very systematic to date, recommendations are provided in Chapter 6 for utilizing spiral development.

• *Review of important questions that remain for experimentation and how they should be addressed.* Chapter 5 evaluates the adequacy of the naval experimentation program, processes, and results. It lists specific areas for experimentation that are not in the current program but that are required for the development of concepts, such as that of network-centric operations, or to support already-programmed capabilities. Chapter 6 expands on recommended areas for future experimentation.

• *Review of needed process and method improvements.* The NWDC's processes for planning and accomplishing FBEs are described in Chapter 3, and its approaches to identifying and selecting key concepts and developing them through experimentation are elaborated in Chapter 5. The MCCDC's processes for experimentation, including concept development, are discussed in Chapter 3. All of the key naval experimentation processes are assessed in Chapter 5, and some specific shortfalls are mentioned. Chapter 6 provides recommendations for improving processes and methods.

• *Review of environment and tools for experimentation.* Chapter 2 defines and delineates an environment and a set of tools to support experimentation. The environment and tools currently in use by naval experimenters are assessed in Chapter 5, and certain shortfalls are noted. Some, but not all, of the items noted will be required for Service-unique experimentation as well as for joint experimentation in the future, according to the planned experimentation campaigns of USJFCOM discussed in Chapter 4. Chapter 6 provides recommendations for addressing the shortfalls.

• *Review of joint experimentation and its relationship to Service experiments.* Joint experimentation has been evolving over the past several years and even during the course of this study. Chapter 4 discusses joint experimentation from three perspectives: that of the U.S. Joint Forces Command, of the Regional Combatant Commanders, and of cross-Service experimentation. It then explores the implications of naval linkages to joint experimentation. Chapter 5 assesses the relationship of Service-unique and joint experimentation with respect to planning, processes, programs, and tools. Areas of progress are noted, as are areas for improvement. Chapter 6 elaborates on the committee's recommendations for improvements—for example, in suggested guidelines for balancing joint and Service-unique experimentation.

• *Review of Service experimentation programs in preparing for joint operations.* The committee reviewed Service experimentation programs as a preparation for joint operations through the perspective of balance—how to maximize the Service participation in and influence on joint experimentation in the future while evolving core Service capabilities that are also a key to joint operations. This area is addressed in Chapter 5, where the need for alignment and synchronization between Service and joint experimentation is noted. Chapter 6 provides recommendations for achieving the desired balance, as well as ways to maximize participation in joint experimentation as a precursor to improved joint operations.

2

Experimentation—What It Means

The Navy and the Marine Corps have embraced experimentation as a fundamental tool for force development. This chapter examines what it means to experiment in the context of military operations. In addition, it explores the methodology of experimentation and experimentation campaigns, including the application of spiral development. It also discusses the environment necessary to support a sound experimentation program. The chapter ends with a discussion of a few practical considerations for military experimentation.

WHAT IS AN EXPERIMENT?

Committee's Working Definition

For the purposes of this report, the committee chose a relatively broad definition of the term "experimentation" in the military context: *Military experimentation is a military activity conducted to discover, test, demonstrate, or explore future military concepts, organizations, and equipment and the interplay among them, using a combination of actual, simulated, and surrogate forces and equipment.*

The definition highlights the fact that building future naval forces through experimentation means more than acquiring the equipment of the future. Building tomorrow's Navy and Marine Corps means developing the doctrine, organization, training, materiel, leadership, personnel, and facilities (DOTMLPF) that together constitute the mission capability of a military force. If experimentation is to be useful, it must deal with all these elements of capability.

Naval experimentation serves a purpose. It is not an end unto itself, nor is it merely a means for pursuing a few interesting ideas. It is intended to build future naval capabilities and must be so crafted. Experimentation must support learning what needs to be known by exploring the potential value of new systems and new ways of operating forces, so that leaders can make informed decisions about advancing the capabilities of tomorrow's naval forces.

The committee recognizes that other definitions of experimentation are in use and plausible. Appendix C provides definitions of experimentation terms used in this report.

What Others Say an Experiment Is

There is surprisingly little agreement among definitions of the term "experimentation" in the military context. Joint Chiefs of Staff (JCS) Publication 1, the DOD's authoritative source for standard definitions, does not define it. The U.S. Joint Forces Command's glossary defines "joint experimentation" as the "application of scientific experimentation procedures to assess the effectiveness of proposed (hypothesized) joint warfighting concept elements to ascertain whether elements of a joint warfighting concept cause changes in military effectiveness."[1] Given actual practice, this definition is relatively narrow with respect to both the purpose of joint experimentation and its methods. U.S. congressional defense appropriation language calls for joint warfighting experimentation to be carried out "in field environments under realistic conditions against the full range of future challenges. . . ."[2]

In *Code of Best Practice for Experimentation*, co-author David Alberts, director of research and strategic planning of the Office of the Assistant Secretary of Defense, distinguishes among three types of experiments, each based on a separate meaning of the word "experiment" and each having a distinct purpose:

- *Discovery experiments* are conducted "to determine the efficacy of something previously untried."[3] These experiments are similar in purpose to joint experimentation as defined in the USJFCOM glossary.

---

[1]See the Web site <www.USJFCOM.mil/about/glossary.htm>. Accessed August 8, 2003.

[2]Strom Thurmond National Defense Authorization Act for Fiscal Year 1999, P.L. 105-261, 112 Stat. 1920, Sec. 921 (October 17, 1998).

[3]David S. Alberts, Richard E. Hayes, John E. Kirzl, Leedom K. Dennis, and Daniel T. Maxwell. 2002. *Code of Best Practice Experimentation,* DOD Command and Control Research Program, Office of the Assistant Secretary of Defense (Networks and Information Integration), Washington, D.C., July, Ch. 3., p. 3-1. Available online at <http://www.dodccrp.org/>. Accessed October 7, 2003.

- *Hypothesis-testing experiments* are used to advance knowledge by "seeking to falsify specific hypotheses."[4]
- *Demonstration experiments* recreate known truth to "display existing knowledge to people unfamiliar with it."[5]

As Alberts points out, the DOD and the Services conduct experiments of all three types. Congressional directives favor Alberts's first two types over his third. The Defense Authorization Act of FY1999 stipulated that joint experimentation was to "investigate and test technologies and alternative forces and concepts."[6] More specific goals were provided in the Defense Authorization Act of FY2000—for example, improving interoperability, synchronizing technology fielding, and developing joint operational concepts.[7] The law is silent on the question of demonstrating capabilities that are already understood for the purposes of achieving buy-in or sustaining momentum.

In the 1990s, the U.S. Army dubbed its Force XXI experimentation series advanced warfighting experiments (AWEs). The term was subsequently adopted by Congress and by others in the DOD. More recently, the DOD has favored the term "field experiments." The USJFCOM glossary defines "field experiments" as "wargames conducted in the actual environment with military units and equipment." It does not offer a separate definition for "warfighting experiments."[8] The 2001 Quadrennial Defense Review refers to "field exercises that incorporate experimentation."[9] Both of these sources hold that such exercises should emphasize the operational level of war rather than the tactical or strategic. Because the term "field experiment" is used today to mean the same thing that "warfighting experiment" meant 4 years ago, the committee has chosen to use the two terms interchangeably.

---

[4]David S. Alberts, Richard E. Hayes, John E. Kirzl, Leedom K. Dennis, and Daniel T. Maxwell. 2002. *Code of Best Practice Experimentation,* DOD Command and Control Research Program, Office of the Assistant Secretary of Defense (Networks and Information Integration), Washington, D.C., July, Ch. 3., p. 3-3. Available online at <http://www.dodccrp.org/>. Accessed October 7, 2003.

[5]David S. Alberts, Richard E. Hayes, John E. Kirzl, Leedom K. Dennis, and Daniel T. Maxwell. 2002. *Code of Best Practice Experimentation,* DOD Command and Control Research Program, Office of the Assistant Secretary of Defense (Networks and Information Integration), Washington, D.C., July, Ch. 3., p. 3-4. Available online at <http://www.dodccrp.org/>. Accessed October 7, 2003.

[6]Strom Thurmond National Defense Authorization Act for Fiscal Year 1999, P.L. 105-261, 112 Stat. 1920, Sec. 921 (October 17, 1998).

[7]National Defense Authorization Act for Fiscal Year 2000, P.L. 106-65, 113 Stat. 512 (October 5, 1999).

[8]See Web Site <http://www.USJFCOM.mil/about/glossary.htm>. Accessed August 8, 2003.

[9]Donald H. Rumsfeld, Secretary of Defense. 2001. *Quadrennial Defense Review Report,* Washington, D.C., September 30, p. 36.

The committee settled on its relatively broad definition, which includes discovery and exploration, test, and demonstration[10]—all three of Dr. Alberts's purposes—and espouses a relatively relaxed view of the scientific rigor with which experiments must be conducted.

Experimentation Includes a Spectrum of Activities

Experimentation is not limited to live events. Rather, it requires a spectrum of activities to advance the understanding of future concepts, organizations, and equipment and to demonstrate their benefits. These activities include studies and analyses, seminars and conferences, work by subject-matter experts, systematic interviewing of experienced officers, war games, and modeling and simulations, as well as live events in the field.

To amplify this point: *The term "experimentation" as used in this study implies all activities in this spectrum, not just field experiments.*

Figure 2.1 illustrates this spectrum. Any of the activities along the spectrum can be repeated at increasing levels of resolution as knowledge grows and systems and concepts take shape (hence the multiple boxes for each type of activity in the figure). Experimentation relies on studies and analyses as overarching activities to be conducted throughout the process, not just employed at the conclusion to explain the results of an individual live experiment. Studies and analyses include the development of theory and concepts to make sense of the whole; systems analysis and systems engineering for understanding systems of systems; empirical analysis to convert data into knowledge; and, ultimately, policy analyses to help decision makers choose among competing options.

Even field experiments, depicted in Figure 2.1 as a discrete component of experimentation, engage supporting activities. Not every experiment requires live play, wholly or even partially by fleet forces, although typically some level of forces is engaged. Even quite large experiments can be conducted using linked models and simulations.

Experimentation Campaigns Advance Understanding Systematically

Using experimentation for selecting, developing, and implementing future capabilities means invoking a full spectrum of activities systematically.

An experimentation campaign is a planned and cohesive, multiyear program of experimentation built on a series of experiments and related activities to

---

[10]It seems to be a common impression that "demonstrations" are biased against failures, in contrast to "experiments," which allow failures. However, demonstrations have been part of the Navy's fleet battle experiments, and some of these were not entirely successful. Demonstrations with mixed results have also occurred in experiments by the other Services and by the Department of Defense.

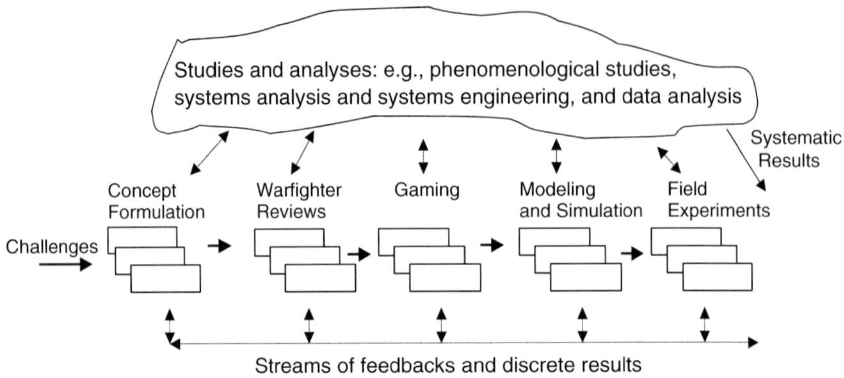

FIGURE 2.1 Spectrum of experimentation activities (see discussion in text).

develop the knowledge needed to inform major decisions about future forces, explore the viability of potential or planned changes to forces or their capabilities, and/or confirm that planned development and directions will lead to capabilities that perform as expected.

An experimental campaign may be constructed to answer questions about one or several elements of DOTMLPF. It should be in alignment with the future vision of the Services and joint forces. The spectrum of activities includes seminars, work by subject-matter experts, studies and analyses, war games, and modeling and simulations, all of which usually precede and support actual field experiments. The field experiments themselves have varying objectives and sizes. A campaign typically uses smaller experiments that incrementally build to larger ones.[11] A campaign must be flexible and may include the iteration of steps, regress to earlier stages, or skip a stage as results warrant. In fact, a campaign plan may be viewed as a "living document" subject to continual replanning.

---

[11]The size of various Service and joint experiments varies greatly: some experiments involve only a small number of lower-echelon units, whereas those conducted at the Joint Task Force level involve thousands of troops from all four Services. Both the Navy and the Marine Corps as well as the other Services conduct limited-objective experiments (LOEs), which focus on one or a few new capabilities in a single area. LOEs can cut across Services and geographic locations and can even include multinational involvement while remaining relatively small in scale. The Marine Corps makes extensive use of limited technical assessments (LTAs), experiments conducted to test the utility of a single technology or piece of equipment in some range of future tasks. By contrast, the Navy's FBE series aims to explore multiple concepts simultaneously and to involve a substantial portion of a fleet, usually while the fleet is engaged in training exercises.

To achieve objectives that embody dramatic departures from current capabilities, experimentation must be carried on continuously and systematically. No single experimentation activity is conclusive. Each experiment uses a sample of the surrounding environment that affects its outcome. The environments and scenarios adopted for an individual experiment are chosen from a range—often a very wide range—of possibilities. Generalizing to the numerous situations in which the findings may be applied is always risky. Experimentation campaigns reduce risk by addressing a spectrum of possibilities and build on successive activities systematically, with analyses done at every step.

A well-planned, prioritized experimentation campaign provides a framework for learning much of what needs to be known about new capabilities: whether they are desirable and feasible, how they compare with other options, and what risks will be involved in developing and fielding them. Campaigns can involve the validation and verification of models for such things as computer applications and the dynamic behavior of planners and resource managers. Structuring a campaign allows planners to proceed along multiple axes of investigation while organizing events around broad goals and objectives. It also introduces multiple decision points, both for experiment planning and for identifying and prioritizing the interlinked changes in forces, equipment, concepts, and organizations that are the main objective of an experimentation campaign.[12] It is important to understand why certain aspects of experimentation activities fail to meet expectations and how things might work under different conditions, and to document these results and apply them in subsequent experimentation.

Thus, a well-structured campaign enables the knowledge gained from each experimentation activity to support and shape the succeeding activities. A campaign is definitive enough at each major step to inform decisions about future research and technology programs, acquisition efforts, organizational changes, and changes in operational concepts.

BUILDING CAPABILITIES THROUGH EXPERIMENTATION CAMPAIGNS

As noted above, individual experiments do not provide the answer to every capability question. In fact, they can result in false impressions about the desirability or feasibility of changes in fighting concepts, force structure, organizations, and equipment. This section highlights some of the limitations of relying too heavily on individual experiments or on simple threads of experimental activities and discusses how well-structured campaigns can mitigate these limita-

[12]David S. Alberts, Richard E. Hayes, John E. Kirzl, Leedom K. Dennis, and Daniel T. Maxwell. 2002. *Code of Best Practice Experimentation,* DOD Command and Control Research Program, Office of the Assistant Secretary of Defense (Networks and Information Integration), Washington, D.C., July, Ch. 3. Available online at <http://www.dodccrp.org/>. Accessed October 7, 2003.

tions. While campaigns cannot explore every scenario, alternative, or excursion, they amplify understanding considerably.

Lack of Scalability

In a world of constrained resources, even a large, live experiment typically involves no more than a small portion of the total force that would ultimately be affected by any of the changes being explored. Thus, the lessons revealed in such a setting may not be scalable to a wider setting. This limitation can be particularly nettlesome for network-centric operations. For example, the performance of a surrogate communications network may support collaboration when only a few users are online, but be so poor as to make collaboration impossible for a larger number of linked operators. Analysts can sometimes rely on models and simulations to extrapolate from experiments to larger settings, but only if the models themselves are based on an understanding of the actual scalability of the variables involved. When cause-and-effect relationships are nonlinear and complex, gaining that basic understanding can require a sustained, iterative program of experimentation. The incremental structuring of activities within an experimentation campaign—such as building from smaller limited-objective experiments (LOEs) to larger FBEs and joint force experiments—can test scalability as well as the ability to integrate the partial results of the smaller experiments into the operation of large-scale forces. Iterative activities can highlight the problems and point toward solutions.

Transferability of Lessons Across the Force

Lessons learned in an experiment about one combination of ships, units, or locations may not apply to the next combination. Even an aggressive and carefully designed experimentation campaign will never be able to illuminate every such combination. As with scalability problems, a healthy dose of model-based analysis, simulation, and wargaming can help analysts to extrapolate the results from a few combinations to many.

Transferability of Lessons from Experiments to Wartime Operations

Safety concerns, resource constraints, artificially benign environments, and the absence of a real adversary make it impossible for experiments to reflect all of the actual conditions of military operations. As a result, lessons learned from experiments apply imperfectly to wartime operations. This limitation can be mitigated by including a strong, motivated, and creative opposing force in the experimentation program, allowing for the exploration of alternative paths when things do not go as planned. This limitation can also be mitigated through the modeling and simulation of various situations and degraded conditions.

The Human Factor

Expanding the knowledge base by using experiments that involve people working together and making decisions in complex new situations will never be easy. As David Alberts points out,

> The variety of applicable military contexts and the rich variety of human behavior and cognition argue for care. . . . Many innovations currently of interest, such as collaborative work processes and dispersed headquarters, have so many different applications that they must be studied in a variety of contexts. Others have organizational and cultural implications that must be examined in coalition, interagency, and international contexts.[13]

One problem in this as in any endeavor is that people can make mistakes. Human errors may be recognized or not. Recognized errors typically do not cause serious problems because the data that they affect can be analyzed separately or discarded. However, mistakes that go unrecognized can lead to false interpretations that destroy the validity of an experiment.

Another set of problems can be introduced when people play roles in an experiment. Research indicates that when complex decision making is involved, there may be differences in performance between people who hold the roles in reality and those who play their roles during experimental events. Bringing actual staffs into experiments obviates those differences, but using actual staffs is not always possible. When role-playing is involved, experiment planners and evaluators need to consider how such differences in decision-making style and skill might affect experimental outcomes.

More generally, different people respond to the same situation in different ways. Published job performance experiments by psychologists show large variations in performance among individuals conducting the same task.[14] Furthermore, the same person may respond differently at different times, depending on training, experience in an experiment, or level of fatigue or stress. The effects of this limitation can be reduced by selecting the experimental operators randomly, including a large number of operators, establishing baseline and control groups, controlling for learning effects during experimentation, and training operators to uniform standards before an experimental event. In addition, human factors

[13]David S. Alberts, Richard E. Hayes, John E. Kirzl, Leedom K. Dennis, and Daniel T. Maxwell. 2002. *Code of Best Practice Experimentation,* DOD Command and Control Research Program, Office of the Assistant Secretary of Defense (Networks and Information Integration), Washington, D.C., July, Ch. 3., pp. 3-10 and 3-11. Available online at <http://www.dodccrp.org/>. Accessed October 7, 2003.

[14]For example, Frank L. Schmidt and John E. Hunter, 1983, "Individual Differences in Productivity: An Empirical Test of Estimates Derived from Studies of Selections Utility," *Journal of Applied Psychology*, Vol. 68, No. 3, August, pp. 407-414, describes variations of two to one or higher in the ratio of productivity between 95th- and 5th-percentile performers.

analysis can help to illuminate the differences among the operators who serve as experimental subjects, and modeling and simulation can improve understanding of how those differences might affect operational outcomes.

All of the potential problems described here argue for conducting multiple rounds of experiments to mitigate some of the limitations caused by complexity and the human factor.[15]

Unrealistic Surrogates

Because the equipment of the future has not been built yet, experiments typically must rely on commercially available surrogates or legacy systems modified to emulate the expected performance of postulated new systems. Such surrogates are needed to help people flesh out new operational concepts and to allow operators to explore the new concepts. However, they may be far from representing the actual capabilities that would be required in real future operations. For example, a surrogate may lack the physical hardening, the information security, the actual performance, or the capacity that would be needed in a fight against a real enemy. To explore the desirability of new concepts under such circumstances, experimenters typically tailor the experiment—for example, by constraining the conditions under which the opposing force can realistically operate. In the heat of enthusiasm, however, it is easy for both experimenters and decision makers to lose sight of the fact that surrogates, while necessary to carry out an experiment, in fact are poor substitutes for the equipment that would be needed to make the concepts real. As a result, experiments can lead to misunderstandings about how difficult it might or might not be in the future to acquire the new capabilities that the Navy and the Marine Corps are considering. Repeated, unvarnished discussion of the role and limitations of the surrogates by experimenters and advocates may help mitigate this class of limitations. Close coupling between the experimentation program and the research and technology base might also help.

Experimentation Campaigns in Summary

Single experiments are insufficient to provide the understanding needed to advance naval forces and their capabilities. Experimentation campaigns enable multiple axes of investigation, with a spectrum of activities that can be matched to the question posed. While no amount of foresight will prevent every surprise, a well-designed experimentation campaign can mitigate the limitations of single

---

[15]David S. Alberts, Richard E. Hayes, John E. Kirzl, Leedom K. Dennis, and Daniel T. Maxwell. 2002. *Code of Best Practice Experimentation,* DOD Command and Control Research Program, Office of the Assistant Secretary of Defense (Networks and Information Integration), Washington, D.C., July, Ch. 3. Available online at <http://www.dodccrp.org/>. Accessed October 7, 2003.

experiments and simple threads of experimental events by combining repeated and persistent live experiments with modeling, analysis, wargaming, and other appropriate activities, building systematically upon the knowledge derived from each activity. *Consequently, experimentation campaigns, not just experiments, are essential for building future naval forces.*

SPIRALING IN EXPERIMENTATION

In this report the term "spiral" is applied to processes in four different contexts related to experimentation. Box 2.1 summarizes these briefly. This section discusses the importance of the well-known spiral development process and expands on spiral processes in experimentation campaigns.

Spiral Development

In 2001 and 2002, the DOD revised its acquisition directives to specify evolutionary acquisition as the preferred strategy for rapid acquisition of mature technology and spiral development as the means to implement it.[16]

Spiral development makes it possible to accelerate the delivery of capabilities that cut across multiple user communities, that involve rapidly changing technologies, and/or that have initially vague or uncertain requirements. This advantage makes spiral development promising for network-centric naval forces and is one of the reasons it was prescribed in the Sea Trial process[17] and recommended by the Naval Studies Board's report on network-centric naval forces.[18] For capabilities that depend heavily on rapidly advancing computing and communications technologies, spiral development is essential if the military forces are not to lag adversaries that can acquire commercially available technological advances. Spiral development is not the preferred choice for every acquisition. For example, in the development of a new ship propulsion system, work on the initial concept and analyses of requirements that match user needs would likely benefit from a spiral process. But the detailed engineering design work entailed in

---

[16]DOD 5000 Series Resource Center. 2002. Department of Defense DOD Instruction 5000.2, "Attachment 2, Operation of the Defense Acquisition System," Defense Acquisition University, Fort Belvoir, Va., October 30, Section 3.3. Available online at <http://dod5000.dau.mil/Memo50002Oct30.doc>. Accessed October 8, 2003.

[17]"Embracing spiral development, these technologies and concepts will then be matured through targeted investment and guided through a process of rapid prototyping and fleet experimentation." ADM Vern Clark, USN. 2002. "Sea Power 21: Projecting Decisive Joint Capabilities," *U.S. Naval Institute Proceedings*, Vol. 128, No. 10, October, pp. 32-41.

[18]Naval Studies Board, National Research Council. 2000. *Network-Centric Naval Forces: A Transition Strategy for Enhancing Operational Capabilities*, National Academy Press, Washington, D.C., p. 78.

BOX 2.1
Spiral Processes in Four Contexts

The term "spiral" is applied to processes used in four different contexts in this report. Each form is applicable to experimentation.

- *Spiral exploration of broad concepts.* A process that, within a given phase of inquiry, uses a variety of instruments (e.g., models, games, and experiments) to explore concepts broadly and uses the same instruments to adjust and iterate until a refined understanding of the issues has been obtained.
- *Spiraling in an individual experiment.* A process that uses many experimental instruments (e.g., games, simulations, and focused experiments) sequentially in preparation for a subsequent complex individual experiment.
- *Spiraling in an experimentation campaign.* A process that uses a spectrum of experimental events to investigate a complex problem incrementally; each spiral typically involves many events and complex experiments, and each subsequent spiral addresses problem complexity that is greater than and/or different from what preceded it (e.g., when new operational concepts for ground forces are studied at successive levels—company, then battalion, then brigade).
- *Spiral development.* A process of evolutionary acquisition that iteratively develops a defined set of capabilities within each increment, providing the opportunity for interaction between the user, tester, and developer; refines requirements through experimentation and risk management; allows for continuous feedback; and provides the user with the best possible capability within each increment.[1]

---

[1]After several years in which multiple offices and organizations adopted their own definitions of spiral development, the acquisition leadership within DOD stipulated a single definition for both spiral development and evolutionary acquisition, which spiral development is meant to implement. According to this DOD definition, spiral development is "an iterative process for developing a defined set of capabilities within one increment. This process provides the opportunity for interaction between the user, tester, and developer. In this process, the requirements are refined through experimentation and risk management, there is continuous feedback, and the user is provided the best possible capability within the increment. Each increment may include a number of spirals. Spiral development implements evolutionary acquisition." Quoted from memorandum from Under Secretary of Defense E.C. Aldridge, Jr., to the secretaries of the military departments and others, dated April 12, 2002, p. 2.

advancing immature technologies and mitigating technology and cost risk may be better handled through more traditional, deliberate development. Propulsion power-source design, such as new generations of gas turbines or turboelectric drives, involves the confluence of many technological advances—in materials, fluid flow management, controls, modes of transmitting power, and propulsion design (noncavitating propellers or water jets, for example)—all of which move ahead at different rates and achieve progress milestones at different times. Thus, for some acquisitions, longer development times and major block enhancements may be the preferred method.

Spiraling in an Experimentation Campaign

By organizing a campaign of experimental activities into spirals, planners can ensure that integrated capabilities are explored at each step so that a coherent set of military capabilities is grown rapidly over a short period of time. With this approach, which works well in procuring large and complex systems and systems of systems, early expression of technical requirements is important. Every effort should be made to design the desired system correctly up front and to make it evolvable, with adaptations then achieved through spiraling. Often, "straw" (first approximation) requirements and designs used at the outset of a campaign are themselves the outcome of earlier experimentation.

Typically, the spirals within a campaign involve increasingly complex, individual live or simulated events aimed at gaining a refined understanding of the concepts and capabilities under consideration. Spirals may also be used to explore the sorts of scalability issues involved in moving from smaller to larger units.[19] Alternatively, spirals may be used to break a broad area of inquiry or a complex set of decisions into more manageable chunks, with detailed exploration of multiple facets building over time to a broader set of integrated capabilities.

The spirals in a campaign are not simply a sequence of events. Rather, they support a continuous and integrated feedback process that engages all of the stakeholders (operators, developers, testers, architects, and others), with an analysis of risks and capabilities and with minispirals to facilitate backtracking in order to refine understanding and resolve issues identified during successive cycles. The spirals are undertaken with the intent, for example, of transitioning the resulting new equipment or systems into service if the outcome is successful, or of actually changing the force's operating procedures; such a tangible outcome is the reason for the expense and effort attending the experimentation campaign.

Figure 2.2 illustrates an experimentation campaign that coevolves a mission capability package beginning with a straw DOTMLPF, which itself may have resulted from prior experimentation activities. Each campaign spiral is directed toward its own specific set of objectives, which are related to decisions about future forces. Each spiral requires its own system-of-systems architecture and integration efforts, which in turn are refined from spiral to spiral as experimental components of DOTMLPF grow in complexity. Within each campaign spiral, end-to-end testing and dry runs are conducted before major live events or simula-

---

[19]For example, the U.S. Army structured its experiments related to digitization of the battlefield to explore issues first at the company and battalion levels, then at the brigade level, and later at the division level.

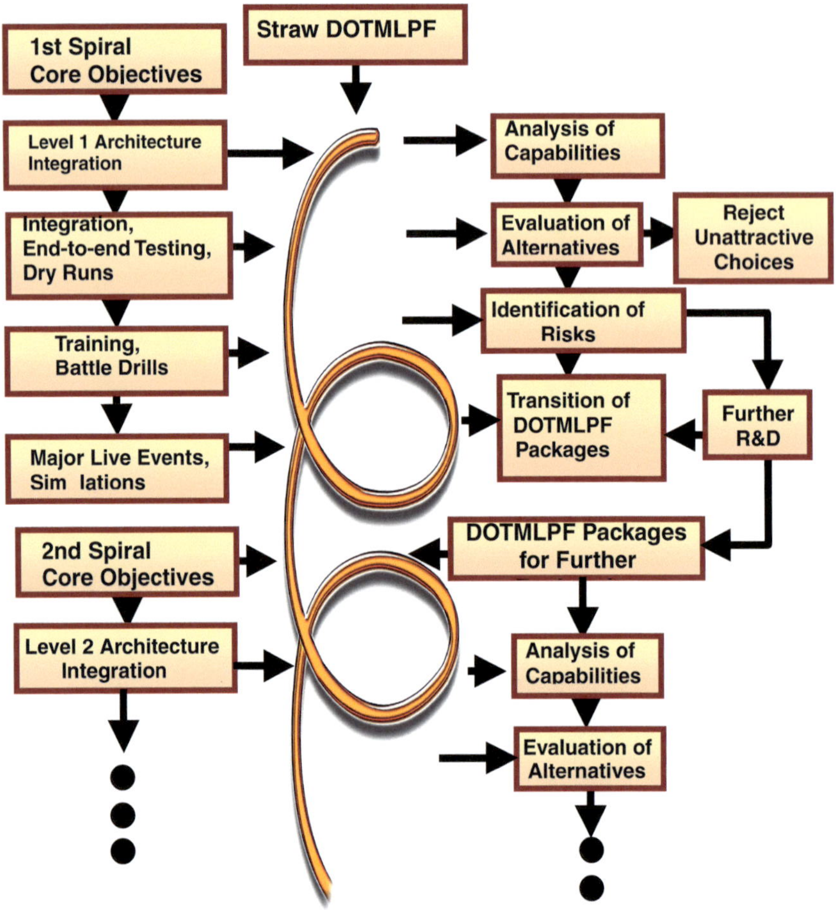

FIGURE 2.2 Spiraling in an experimentation campaign.

tions in order to bring to light problems in concepts, systems, organizations, doctrine, TTPs, and training.

In the campaign illustrated, the spirals and their culminating events should be defined so as to allow for rapid incremental development and, if the development is successful, for the folding in of the new capabilities—thus linking firmly into the spiral development/evolutionary acquisition model noted earlier. At culminating points throughout the campaign, mature, verified, and tested capability packages should be transitioned into new, fielded capabilities. Some packages may require new acquisition program starts; others may modify existing programs of record. Still others may require no changes in material but may instead involve changes in other elements of DOTMLPF. The less-mature concepts, organizational changes, and technologies are either set aside or moved into the next spiral of experimentation.

THE ENVIRONMENT FOR EXPERIMENTATION

As noted earlier, experimentation relies on a wide spectrum of experimental activities, from studies and analyses, seminars and conferences, war games, modeling and simulation, to live play events. Deriving the maximum benefit from experimentation requires an environment that supports all such activities.

An experimentation environment, as illustrated in Figure 2.3, provides a culture, committed leaders, and skilled personnel as well as infrastructure and tools. Box 2.2 expands on representative elements of such an environment, and these are briefly discussed in subsequent paragraphs.

Leadership and a Culture of Learning

Experimentation is naturally disruptive of the status quo. For experimentation to contribute to building future forces, senior leadership must support a free play of ideas that run counter to traditional culture and expectations. Because learning about ideas that do not succeed can be as important as learning about those that do, experimentation requires a culture that accepts negative results and occasional failures, coupled with an understanding that the learning itself is a successful outcome. A culture that rewards risk takers and encourages innovative behavior is key to capitalizing on experimentation. Equally important is the empowering of subordinates to make appropriate decisions, to work creatively, and to put forward ideas that may be unconventional and disruptive.

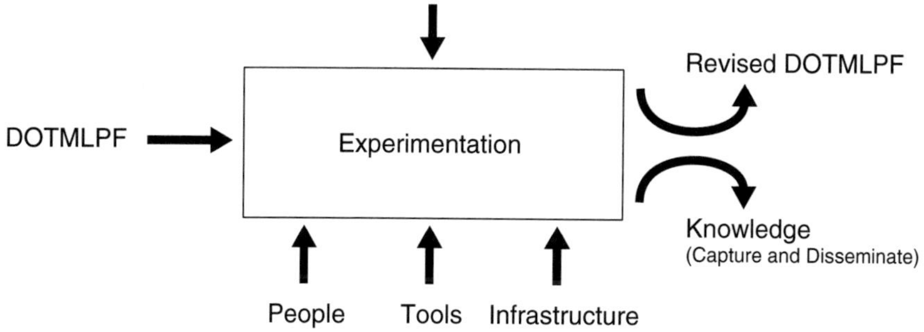

FIGURE 2.3 Experimentation environment. (See Box 2.2 for expanded list of elements.)

BOX 2.2
Major Elements of Experimentation Environment

Leadership and Culture
- Committed leaders
- Cultures of learning
- Incentives for risk
- Tolerance of negative results
- Empowerment

Information and Physical Infrastructure
- Networks
- Information repositories
- Architectural frameworks
- Integration and test facilities
- Training facilities
- Places and platforms

Trained and Talented Personnel
- Concept developers
- Systems analysts
- Operators
- Red-team cells
- Support teams

Tools
- Modeling and simulation
- Prototypes, surrogates, and so on
- Artificial environments
- Data capture and dissemination

Trained and Talented Personnel

Experimentation requires a commitment to providing talented personnel who are dedicated to experimentation. Each member of the experimentation cadre must be both educated and experienced in experimentation, in addition to having a discipline related to warfare.[20] The mix of expertise required in any one experimental activity may vary depending on the nature of the activity, but core expertise includes concept developers, system analysts, operators, an opposing-force cell, and a support team of engineers, developers, testers, and trainers. Capable concept developers are essential, because without innovative new concepts to explore there can be no experimentation of a far-reaching nature. A substantive understanding of military operations plus a good imagination is required of these individuals. Systems analysts, typically working with the aid of models, are

[20]According to David Alberts, "The single most important consideration for those responsible for experimentation design, whether single experiments or campaigns, is to ensure current expertise is available to support the plan." See David S. Alberts, Richard E. Hayes, John E. Kirzl, Leedom K. Dennis, and Daniel T. Maxwell, 2002, *Code of Best Practice Experimentation,* DOD Command and Control Research Program, Office of the Assistant Secretary of Defense (Networks and Information Integration), Washington, D.C., July, Ch. 6. Available online at <http://www.dodccrp.org/>. Accessed October 7, 2003.

required to assess in detail the operational and technical feasibility of the concepts being developed.

When the concept under consideration has reached a degree of maturity, the experimentation cadre will design and execute experimentation activities to test it. This cadre should have a broad range of both operational and technical expertise—operational, for example, in order to develop scenarios and determine data collection needs, and technical to understand the necessary simulation capabilities and systems integration. General logistical support in experiment preparation and execution is also required. Analysts are also critically needed for planning, designing, and evaluating experimentation events.

Operators—i.e., the actual "players" in the experiment—must be familiar with the concept (or capability) being explored and with the use of the tools (e.g., simulations) being employed and must also be of a mind-set to explore and innovate. Playing against the operators will be an opposing force, or red-team cell. This independent body is necessary for exposing, in as honest and probing a manner as possible, potential flaws in the concept being explored.

Information and Physical Infrastructure

Most experiments will require network connectivity and capacity as well as computing power and data storage. In addition, they depend on the existence of accepted architectural frameworks (e.g., for simulation, the High-Level Architecture—Institute of Electrical and Electronics Engineers standards P1516, P1516.1, and P1516.2) for integrating simulation assets. The experiments not only will generate data but also will require much data, such as scenarios to drive the experimentation play. Having repositories of such information available and accessible can ease experiment planning and enable the extension of results to cases or scenarios that are not included in the live play.

Ultimately, a system that links platforms into a virtual environment is needed, to provide an appropriate level of realism in live experimentation conducted across a fleet. Such a system supports the controlled insertion of simulated threats and other scenario elements into the command and control environment. Embedding the system directly into training systems aboard each platform may help naval operators shift seamlessly between training and experimentation.

An experiment and any or all related activities must take place somewhere. Key facilities, such as those for simulations, for war games, for integration and testing, and for training, must be provided. Such facilities must be equipped with the basics of uninterrupted power, good lighting and ventilation, and suitable climate control, as well as with the space and support for any specialized equipment needed for specific activities. Equally critical is the timely availability of ships, test ranges, aircraft, and various platforms that are integral to the experiments—this need is particularly challenging for naval experimentation. Integration facilities and environments are critical because integration of needed components to support experimentation is challenging,[21] often encompassing systems, databases, sensors and weapons platforms, and networks, as well as various tools to support experimentation. Interoperability is often difficult to achieve, exacerbated by the relative immaturity of the concepts and capabilities that are being explored.

Tools

Many other types of tools are required to support experimentation, such as those for establishing artificial environments and for supporting actual prototypes and managing the use of surrogates. The interoperability and integration of tools are often challenging, requiring a defined framework and an appropriate integration environment, as noted above in the discussion on information infrastructure.

Models and simulations play an essential role in experimentation, given that physical platforms and facilities may not always be available or in fact may not yet exist. The models and simulations can range from largely computer-driven constructive models to virtual simulations in which human participants in platform simulators or at command-and-control terminals make real-time battle decisions.[22] The constructive models can be of varying levels of resolution—for example, highly aggregated in support of a strategic-level war game, to quite detailed for examining urban operations. Ideally the models and simulations should form an interrelated family to serve and allow ready transitions among the different purposes (e.g., from analysis to wargaming to human-in-the-loop operations).

Several capabilities involved in experiments will be modeled or simulated individually and then linked with the larger models and simulations representing the overall military capability in the experiment. These include prototypes of new systems, surrogates needed because of a lack of availability of some systems, simulations, and artificial or model environments.

Experimentation activities can produce very large volumes of data. For example, a large virtual simulation can involve several thousand entities, with state data (position, velocity, and so on) on each entity being generated on the order of every second. Thus, mechanisms for data collection ranging from automated capture to human observers are necessary. In addition, sophisticated analysis

[21]For instance, see BG Steven Boutelle, USA, and Alfred Grasso, 1998, "A Case Study: The Central Technical Support Facility," *Army RD&A* (now *Army Acquisition, Logistics & Technology (AL&T) Magazine),* March-April, pp. 30-33; and Annette J. Krygiel, 1999, *Behind the Wizard's Curtain: An Integration Environment for a System of Systems,* National Defense University Press, Fort L.J. McNair, Washington, D.C., July.

[22]Constructive experiments use simulated forces in a simulated environment; virtual experiments use partial real forces in a simulated environment; field experiments use real forces in an actual environment.

tools to aid analysts in interpreting the data are required, as are means to store and disseminate the information so derived.

PRACTICAL CONSIDERATIONS IN EXPERIMENTATION

As with other military activities, military experiments take place in the real world. This section discusses the practicalities of dealing with resource constraints and moderating the effects of less-disciplined methods of experimentation.

Contention for Resources

The most onerous practicalities for experimentation are those related to resources. In the U.S. military, money and personnel are typically spoken for years in advance. Dedicated funding for experimentation is essential, given the long lead times required for accomplishing not only campaigns but also individual experiments. Identifying the funding needed to transition the findings from experimentation to the field is very difficult, given the size of the requirement and its unanticipated impact on the programming cycle. Nonetheless, transition funding must be secured if the results of experimentation are to influence and transfer to force capabilities. This issue is addressed in some detail in subsequent chapters, but it should be noted here that the discussion of and contention for resources that accompany any change in a Service's equipment, systems, or mode of operating significantly delay any decision to adopt the results of experiments. The decision cycle must start early. Therefore, in anticipation of success, planning for the transition should start at the same time that planning begins for an experiment or an experimentation campaign. The rationale for this step is elaborated more fully in Chapter 5, in which the progress and problems of Navy and Marine Corps experimentation to date are evaluated.

The competition for resources is exacerbated when the needs of military experiments are overlaid on military training exercises. In the Navy, major events on ships and at air training ranges are typically programmed 6 years in advance. Training schedules for individuals and units are tight even without additional duties. In response to such constraints, the military services and the joint community are increasingly using their training exercises as vehicles for experimentation, in essence tacking experiments onto exercises that units would perform in any case. This occurs so frequently that a military exercise is often characterized as an experiment.

Such opportunistic experimentation allows the Navy and Marine Corps to explore operational-level change that would be nearly impossible to examine using smaller, dedicated forces and allows limited resources, including time, money, and people, to be stretched.

The objective of a military exercise, however, is to ensure and measure today's readiness by training or evaluating military units[23]—quite different from the objective of experimentation, which is to explore and demonstrate future concepts, organizations, and equipment. Piggybacking experimentation on training exercises can mean accepting exercise conditions, forgoing a clear baseline against which to measure change, and limiting the repeatability of events. In some cases, these limitations can be mitigated through careful experiment design. In all cases, setting priorities and doing the best with the resources available are critical. The ramifications of these issues in the naval experimentation environment are discussed in more detail in subsequent chapters.

Overlaying naval experimentation needs with those of joint experimentation can also pose problems, as amplified in Chapter 4. Careful synchronization and alignment of experimental goals, assets, and schedules are necessary to make experiments useful to both the Department of the Navy and the joint community.

Less-Disciplined Field Experiments

Live-play events require greater improved discipline. The various briefings provided to the committee, as well as the views of individual committee members, precipitated a lively debate about the degree to which field experiments must hew to rigorous scientific experimentation procedures if they are to be worthy of the name. In general, those schooled in the physical or biological sciences argue for tighter standards of scientific rigor, including the unambiguous statement of hypotheses, rigorous experimental design, the development of a clear baseline against which to measure change, the establishment of control regimens, careful treatment of independent and dependent variables, repeatability, scalability, and so forth. Experts long engaged in field experiments typically favor more relaxed standards, in large measure because, in their experience, the more rigorous standards seemed unachievable in a field setting, especially for the larger-scale experiments.

If experimentation is to be a credible enabler of future forces, it must adhere to sufficiently rigorous methods. This requirement reinforces the need for a well-planned experimentation campaign that includes an appropriate complement of activities both preceding and following field experiments. Field events should

---

[23]The *Department of Defense Dictionary of Military and Associated Terms* (Joint Publication 1-02) defines an "exercise" as "a military maneuver or simulated wartime operation involving planning, preparation, and execution. It is carried out for the purpose of training and evaluation. . . ." Joint Chiefs of Staff. 2002. *Department of Defense Dictionary of Military and Associated Terms*, The Pentagon, Washington, D.C., April 12 (amended through September 5, 2003), p. 189. Available online at <http://www.dtic.mil/doctrine/jel/new_pubs/jpl_02.pdf>. Accessed October 7, 2003.

continue to improve in rigor; however, it is also important that the full spectrum of activities augment field experiments to ensure the application of good scientific and analytical methods. This combination will convey a more coherent and incisive picture of the results of the fixed experiments, which enables discarding or substantiating and extending findings while determining the many factors that shape these live-play events.

3

Experimentation—Past, Present, and Future

This chapter provides an overview of past, current, and future experimentation programs in the four Services. The emphasis is on naval experimentation, but the experiences and lessons of the other Services are relevant and therefore are included.

U.S. NAVY

The U.S. Navy has a long history and tradition of using experimentation to evaluate doctrine, equipment, and tactics, techniques, and procedures (TTPs)—in fact all of the elements included in DOTMLPF: doctrine, organization, training, materiel, leadership, personnel, and facilities. In October 1884, the Naval War College was established at the behest of Captain Alfred T. Mahan and Rear Admiral Stephen B. Luce, who argued that future naval commanders needed a place where they could develop tactics and doctrine through experimentation. In Mahan and Luce's day, the experimentation consisted of tabletop simulations of fleet maneuvers.

Through the years, the Navy has "experimented" with new platforms (submarines in about 1901, carriers in about 1920, PT boats from about 1939 to 1941) and new propulsion systems and fuels (diesels/fuel oil[1] from about 1904 to 1935). Experimentation conducted from about 1923 to 1940 with exercises that the

[1]See John R. Edwards, 1904, *Report of U.S. Naval "Liquid Fuel" Board of Tests Conducted on the Hohenstein Water Tube Boiler*, U.S. Government Printing Office, Washington, D.C.

Navy termed "Fleet Problems" was key to the development of U.S. carrier doctrine.[2]

A Case History in Past Experimentation: The Early Development of Naval Aviation

The history of the Navy's use of experimentation to achieve new capabilities is illustrated by the role of experimentation in the introduction of aircraft and aircraft carriers. The motivation to undertake an experimentation campaign related to naval aviation was driven directly by a decision of Admiral of the Fleet George Dewey, who began to push the concept of naval aviation after viewing the use of dirigibles. The admiral is said to have commented, "If you can fly higher than the crow's nest, we will use you."[3] To pursue the concept of naval aviation, Captain Washington Chambers was designated by Admiral Dewey as the Navy's lead aviation project officer. Chambers's jobs were to find funding for the project and to demonstrate that an aircraft could both take off from and land on a ship.

George von L. Meyer, then Secretary of the Navy, refused to include funds in the budget for the demonstration. Not to be deterred, Chambers found a rich, politically well connected publisher and aviation enthusiast named John B. Ryan to help him. Ryan contacted President Taft, who persuaded Secretary Meyer to change his mind and designate the cruiser USS *Birmingham* to be used for the experiment. The experiment required the construction of a wooden ramp extending from bridge to bow. While the Navy provided the ship, the cost of the ramp ($288) was paid for by Ryan. The first demonstration of an aircraft taking off from a ship took place near Norfolk, Virginia, in November 1910.

Captain Chambers was then authorized to spend not more than $500 to construct an aircraft recovery ramp on the stern of the cruiser USS *Pennsylvania.* On the basis of experiments ashore, Chambers and his pilot, Eugene Ely, determined that arresting cables would be needed to bring the aircraft to a stop. Accordingly, 15 cables were stretched across the deck, each fastened at either end to a 50-lb sandbag. When the cost of the arresting cables exceeded the funds allocated for the project, Captain C.F. Pond, the skipper of the *Pennsylvania,* paid for the overrun out of his own pocket.

On January 18, 1911, in San Francisco Bay, Ely landed his aircraft on an up-sloping ramp on the rear deck of the *Pennsylvania.* Ely's tail hook caught the 10th arresting cable and his plane stopped 50 ft from a crash barrier. Captain Pond's report after the experiment read:

---

[2]"Fleet Problems" were at-sea exercises with a considerable experimentation component. See Brian McCue, 2002, "Wotan's Workshop: Military Experiments Before the Second World War," Occasional Paper, Center for Naval Analyses Occasional Paper, Alexandria, Va., October.

[3]RADM George van Deurs, USN (retired). 1966. *Wings for the Fleet; A Narrative of Naval Aviation's Early Development, 1910-1916,* U.S. Naval Institute, Annapolis, Md., p. 3.

> This was the most important landing since the dove flew back to the Ark.[4] I desire to place myself on record as positively assured of the importance of the aeroplane in future naval warfare, certainly for scouting purposes. For offensive operations such as bomb throwing, there has as yet, to my knowledge, been no demonstration of value, nor do I think there is likely to be. The extreme accuracy of control as demonstrated by Ely, while perhaps not always to be expected to the same degree, was certainly not accidental and can be repeated and probably very generally approximated to. There only remains the development of the power and endurance of the machine itself, which as with all mechanical things, is bound to come.[5]

In 1913, Captain Chambers determined that all available aircraft and pilots should take part in the fleet's winter exercises of 1913 off Guantanamo Bay, Cuba. These annual exercises were the equivalent of the current Navy fleet battle experiments (FBEs). For these experiments, Chambers's officers rigged a wireless transmitter on one of the aircraft and a receiver on the flagship. An aircraft then flew over the horizon to scout out the position of the opposing forces. Although transmission took place on the plane, reception on the flagship did not occur. However, the concept of using an elevated platform to locate hostile forces had been established.

During the next 7 years, aircraft technology—driven by needs of the Allied and Central powers in World War I—accelerated rapidly, as did the number of qualified flyers and aircraft in the U.S. Navy. By the end of World War I, aircraft carried weapons (machine guns), could drop bombs, and could undertake primitive communications. Experiments had resulted in the development of moderately safe catapults that allowed pontoon aircraft to be launched from a ship's fantail. In 1917, the British Navy undertook experiments with arresting cables that could absorb the energy of a landing aircraft more efficiently than could Ely's arrangement of cables and sandbags. Thus, by the end of World War I, all of the technology required for an aircraft carrier was in place.

On March 20, 1922, the USS *Langley*, the Navy's first aircraft carrier, was commissioned. The ship had been converted from the former *Jupiter*, a collier. By the end of the decade, two more carriers, the *Lexington* and the *Saratoga,* were commissioned. The performance of carrier aviation in the war games (FBEs) of 1929 was a portent of the future. Opposing fleets were charged with the attack and defense of the Panama Canal. The *Saratoga* (attacking force), under cover of darkness and bad weather, launched 69 aircraft, which arrived over and theoretically destroyed the canal without incident. Thus, the role of the fast carrier was predicted 12 years before Pearl Harbor.

---

[4]RADM George van Deurs, USN (retired). 1966. *Wings for the Fleet; A Narrative of Naval Aviation's Early Development, 1910-1916,* U.S. Naval Institute, Annapolis, Md., p. 28.

[5]RADM George van Deurs, USN (retired). 1966. *Wings for the Fleet; A Narrative of Naval Aviation's Early Development, 1910-1916,* U.S. Naval Institute, Annapolis, Md., p. 29.

Evolution of the Linear Development Model

During World War I, Secretary of the Navy Josephus Daniels established a Scientific Advisory Board under Thomas Edison. Edison (over the objections of other board members) pushed for the establishment of a naval experimental station that was to conduct tests and experiments leading to better gun barrels, improved radio communication techniques, improved steel armor, and new torpedoes and mines. The recommended experimental station was established by 1923 and evolved into the Naval Research Laboratory (NRL), which rapidly began to operate as a scientific laboratory in the way that other board members had recommended.

At the end of World War II, the Navy adopted and institutionalized the concept of Vannevar Bush that the process of change in and growth of naval capabilities was a continuous and unending process. New capabilities would result from a linear process that started with an investment in basic research. The results achieved would then enter into a development process that would eventually transition into a state of sufficient maturity so that the results of the R&D process could be incorporated into systems and equipment that could be procured for fleet use. Once new systems and equipment were delivered to the fleet, naval personnel would devise optimum techniques for their employment.

This linear model resulted in some spectacular successes, yielding new warfighting capabilities that grew out of Navy basic research investments. Among these were the Global Positioning System (GPS), overhead surveillance and target localization capabilities, passive undersea surveillance, high-strength steels for submarine hulls, and phased-array antennas for radar and communications systems.

Nonetheless, the linear acquisition model had several deficiencies. These included long delays in delivering products to operators in the field, products that were technologically obsolete by the time of their introduction, and products that did not perform as advertised. In response, the Under Secretary of Defense for Acquisition and Technology (USDA&T) created the advanced concept technology demonstration (ACTD), which was structured to put mature technology in the hands of operators to address a particular need. The goals of ACTD were to determine where, when, and why a system or technology did or did not work and to allow the operator an opportunity to develop TTPs using the technology. Another response to problems experienced with linear acquisition was the adoption of the spiral development method, discussed in Chapter 2.

The Navy has modified its historical approach to developing potential capabilities for the fleet by incorporating both linear acquisition and spiral development methods. Since the mid 1990s, the Navy has participated in various ACTDs such as Cruise Missile Defense, Phase 1; Extending the Littoral Battlespace; Link 16; and Coastal Area Protection System, to name a few. It has also applied spiral development, as is illustrated by the recent case study that follows.

A Recent Experimentation Program: Advanced Rapid Commercial Off-the-Shelf Insertion

A more recent example of the role of experimentation in fostering new naval capabilities—and one using a spiral development approach—can be found in the Navy's current Advanced Rapid Commercial Off-the-Shelf (COTS) Insertion (ARCI) Program. ARCI was motivated by an analysis of U.S. submarines' operational experience in the early 1990s. Operations conducted against quiet Soviet submarines indicated a loss of acoustic advantage. The tactical and strategic implications of this problem were fully appreciated by the Submarine Type Commanders (Commander Submarine Force, U.S. Atlantic Fleet, (COMSUBLANT) and Commander Submarine Force, U.S. Pacific Fleet (COMSUBPAC)), the Submarine Programs Resource Sponsor (Office of the Chief of Naval Operations (OPNAV) N87)), and Naval Sea Systems Command (NAVSEA) 08 (Naval Reactors) Admiral Bruce DeMars, who was the senior submarine officer in the U.S. Navy. The recognition of the loss of acoustic advantage by the submarine force's senior leadership galvanized a multifaceted response that included the acceleration of the existing sonar acquisition programs.

The first step in solving the acoustic superiority problem was the creation of a special advisory group known as the Submarine Superiority Technical Panel. This panel examined the problem and possible courses of action for regaining acoustic superiority. The panel's recommended approach to quickly (and cheaply) improve submarine sonar processing capability was based on a philosophy of "build, test, build." That is, the submarine community developed and tested capabilities and then integrated and installed them together as a unit of incremental capability commonly called a "block" to achieve an improvement in capability. This process was repeated and when an additional level of capability enhancement was achieved, another block was installed in succeeding submarine developments and overhauls. Thus, capabilities were enhanced through a series of block upgrades, phased in incrementally over time, in a process in which spiral development (develop, test, develop) could occur within individual block upgrades.

New algorithms hosted on COTS processors were first tested in the laboratory against data collected from at-sea controlled experiments and real-world operations. The tests were performed in the laboratory by an independent third party. Once an algorithm was determined to have performed successfully in the laboratory, it was taken to sea and tested in controlled experiments. The governing principles in at-sea testing were that (1) operational testing must be adequate and carried out under realistic conditions, and (2) degraded performance must be understood at a fundamental level. Feedback and analysis from the at-sea experiments were used to modify algorithms and correct deficiencies. The system after modification would then be integrated and would undergo an end-to-end test to ensure that it was working properly.

To transition the successful ARCI software and hardware components, a program was established to deploy successful ARCI products into the submarine fleet as quickly as possible—by means of the block upgrades described above.

Almost all 688 class submarines have been or will be upgraded with ARCI technology.[6] The Seawolf- and Virginia-class programs have also integrated the ARCI approach into their sonar systems. The ARCI approach is also being followed by the antisubmarine warfare community, specifically in the Surface Ship Towed Array and Bow Sonar Acquisition Programs.

In addition to testing new algorithms for detection and classification capability, a parallel effort was set up to develop TTPs as well as decision aids to take advantage of the new ARCI-enabled capabilities. In this regard, littoral conditions in terms of both acoustic conditions and contact density (number of objects acoustically detected per unit area) have been emphasized. The ARCI Program has been executed through a close partnership among NAVSEA, OPNAV, the Program Executive Offices, the Operational Test and Evaluation Force, and the fleet.

Experimentation That Changed Operational Capabilities

A review of successes in naval aviation and the example of the ARCI Program provide some lessons with respect to successful experimentation. If experimentation is to enable changes in fielded capabilities, the following five factors need to be present:

1. *Problem to be solved.* First, a significant problem must exist. In the case of naval aviation, the problem was the development of a naval air capability by other countries and adversaries. In the case of the ARCI Program, the problem was U.S. submarines' loss of acoustic advantage—a compelling need.

2. *Availability of technology.* In the cases of naval aviation and ARCI, technologies appeared that made it possible to experiment with TTPs to improve warfighting capability.

3. *Leadership buy-in.* In the cases discussed, top officers in the Navy—the Fleet Commander and the Submarine Type Commanders, respectively—were committed to change.

4. *Organizational structure conducive to change.* In both of these cases, organizations were created under which the testing of new concepts with new technologies could flourish. Also, these organizations reported directly to senior

[6]The Defense Operational Test and Evaluation's Annual Report for 2002 states that ARCI systems, while being deployed in increasing numbers, have not been adequately tested owing to lack of availability of resources: test platforms (submarines) and time (p. 133). The Secretary of the Navy, George England, has been quoted as having assured DOT&E that deployment ". . .risks were . . . considered acceptable . . . to support our emerging plans in the war on terrorism." (Maline Brown. 2003. "Young Wants Navy to Trim Time, Money Spent on Operational Testing," *Inside the Navy*, Vol. 16, No. 4, January 27, p. 1.)

leadership. In the more recent example of ARCI, the Submarine Development Squadron (SUBDEVRON 12) had the responsibility for testing new concepts and new technologies. The DEVRON reported results directly to COMSUBLANT (relevant information was provided to COMSUBPAC) and to the N7[7] and N8[8] senior submarine officers. Owing to this process, important results were transitioned.

5. *Funding.* In both cases, stable funding was provided to move the experimentation process along, and funding was provided to transition capabilities to the field. Funds were established at the specific direction of senior leadership. The President established funding for the first "carrier" experiment, and the Submarine Type Commanders with N76 and N86 support secured funding for sonar improvements.

The committee found that these five factors for success were and are applicable to successful experimentation programs in the other military departments. The experience is reinforced later in this chapter through details on past and present programs of experimentation by the Services. As seen below, the absence of one of these five factors significantly increases the likelihood of a failure to transition the results of experimentation to a future, fielded military capability. This is true regardless of the identity of the sponsoring entity—whether it is a Type Command in the Navy, the Navy or the Marine Corps, the Department of the Navy or any other military department, or the joint community through the U.S. Joint Forces Command (USJFCOM).

Organizational Roles and Major Participants in Navy Experimentation

The Navy Warfare Development Command (NWDC) was established at Newport, Rhode Island, in 1998 to address the coevolution of Navy concepts and doctrine through experimentation. Its mission as briefed to the committee in July 2002 is this:

- To develop Navy warfighting concepts,
- To conduct concept-based experiments,
- To represent the Navy with joint and Service laboratories and tactical development commands, and
- To be the primary point of contact for naval and joint/combined doctrine and experimentation.[9]

[7]N7 is responsible for setting requirements in the Office of the Chief of Naval Operations; N76 is responsible for undersea warfare requirements in the N7 office.

[8]N8 is responsible for allocating resources in the Office of the Chief of Naval Operations; N86 is responsible for allocating resources to undersea warfare in the N8 office.

[9]RADM Robert Sprigg, USN, Commander, Navy Warfare Development Command, "Navy Experimentation Overview and Progress Summary," presentation to the committee on April 5, 2002.

When NWDC was first established, it reported to the president of the Naval War College. As a result of a reorganization by the Navy in 2002, NWDC now reports to the Commander of the Fleet Forces Command (CFFC). The recent change in reporting relationships is intended to strengthen NWDC's ties to the fleets, facilitating the introduction of new concepts, the harvesting of fleet ideas, and a continuous dialogue with fleet customers to explore the merits of various concepts and the evaluation of operational capabilities.

New concepts can come from any source. For example, the focal point for the Navy Global Hawk concept is at the Navy Unmanned Aerial Vehicles Office (designated as PMA 263) in the Naval Air System Command, Patuxent River, Maryland. In principle, concepts proposed for NWDC's consideration may come from the fleet, from the Senior Steering Group, from ONR (or its contractors), from Navy laboratories and warfare centers, from the results of an ACTD, or even from a commercial contractor. In practice, concepts that have some degree of technical maturity and associated funding are given more attention than those that lack technical maturity and funding. Those concepts endorsed by major commands and/or senior officers are most likely to drive NWDC's efforts. NWDC works with the fleet commands in developing experiments and with MCCDC on naval force efforts involving both the Navy and the Marine Corps.

One of NWDC's principal responsibilities is to plan and coordinate fleet battle experiments. The CNO established the Maritime Battle Center (MBC) in 1998 at NWDC to serve as the single point of contact for FBEs. In this capacity, MBC plans, prepares, conducts, and evaluates FBEs in coordination with many participating organizations. NWDC has the decision authority to run limited-objective experiments (LOEs) that do not need large fleet participation. LOEs cost less than FBEs, but their funding sources are different and the visibility of FBE results is greater. These distinctions can create organizational incentives that may not be in the best interests of the Navy as a whole.

Although MBC has been assigned the role of FBE coordinator, many components of the Navy carry out experimentation on a more or less continuous basis. As noted in the sonar improvements case study, the submarine community has established a dedicated squadron whose entire mission is to undertake experimentation with new tactics, doctrine, and technology, so that new capabilities can achieve rapid introduction into the submarine force. The Navy and Marine Corps have a substantial R&D community to produce new capabilities (platforms, weapons, sensors, communications systems, and so on) that are designed to enhance the warfighting capabilities of naval forces. These new capabilities are "experimented with" by computer simulations, by trial on test and training ranges, through war games, and by employment during fleet deployments. Some naval organizations such as the Third Fleet regard the participation in FBEs to be among their most important missions. In addition to hosting FBEs, the Third Fleet provides support on a continuous basis to Systems Commands (SYSCOMs) to experiment with or observe the value of specific developments that the SYSCOMs are sponsoring.

While ONR's primary mission is to manage and foster R&D efforts within the Department of the Navy, in recent years these efforts have included support of "experimentation." ONR has both provided financial support for NWDC and acquired various platforms (SLICE, high-speed vessel (HSV),[10] and so on) that have been used "experimentally" by fleet forces. Recently organized entities such as the Navy Network Warfare Command (NETWARCOM) have proposed extensive use of computer simulations and experimentation to drive transformation. Finally, the Navy has long supported the OPTEVFOR organization, which practices experimentation in the true academic sense of the word. OPTEVFOR examines the hypothesis that some new platform, item of equipment, or software product will improve naval capabilities and consequently should be procured. Although the results of its efforts are frequently negative, OPTEVFOR routinely supports such operability assessments.

Navy organizations involved in planning for and executing Navy experiments in conjunction with NWDC include the following:

- *Warfare centers of excellence*—for concept development, provision of equipment for experimentation, and evaluation of results;
- *Numbered fleets and the Type Commands*—working either directly with NWDC or through Fleet Forces Command on experimental needs and the provision of platforms and other fleet assets for experiments. For instance, the new Navy Network Warfare Command has a special responsibility to coordinate experimental aspects of information systems and information networks. The other Type Commands also coordinate with NWDC on large experiments and for LOEs not requiring large force elements. They may plan and execute their own, smaller experiments (e.g., the submarine sonar experiments described earlier in this chapter);
- *ONR*—for funding some aspects of NWDC's experimental activities and for providing equipment for FBEs. It was noted at NWDC that, if ONR did not provide equipment for experimentation, the FBEs could become science fairs, with industry sponsors providing the equipment for the Navy experiments;
- *The OPNAV staff, and N7 and N70[11] in particular*—for identifying needs and for using the results of experiments in developing and approving Navy requirements; and

---

[10]SLICE is a new, patented ship technology that enables SWATH (small waterplane area twin hull) ships to operate at higher speeds while retaining their characteristic low motions in a seaway. SLICE is not an acronym. High-speed vessels (HSV) are commercially available, leased by the Navy and the Army, for experimentation purposes.

[11]N70 is responsible for requirements analysis in the N7 office.

• *The Assistant Secretary of the Navy for Research, Development and Acquisition (ASN(RDA)), including the program executive offices and program managers for acquisition programs*—as possible identifiers of experimental needs and as recipients of the results of experiments.

In working with the other Navy organizations, NWDC must use its coordinating skills and abilities to direct and leverage all of the participation it can obtain from organizations and parties that it does not control. For example, NWDC has neither the line item funding to control equipment to be used for experimentation (it relies on ONR) nor the authority to require the N7 staff to listen and act on the results of experiments when dealing with Navy "requirements."

FLEET BATTLE EXPERIMENTS

The Navy uses various types of field events in experimentation. Among the most prominent are FBEs, ACTDs, and LOEs. Fleet battle experiments are the most visible and resource-intensive activities in the spectrum of events comprised by experimentation in the Navy today. This section focuses on FBEs and the processes associated with them and reviews results to date. LOEs are also addressed, particularly in association with specific FBEs.

FBEs are field experiments used to address a variety of objectives. When NWDC formulates a concept and evaluates it through various studies and analyses, war games, and simulations, it employs FBEs to explore the concept and supporting technologies in the fleet to determine whether the concept has merit. As an example, several years ago during a global war game, the Navy explored the use of smaller, high-speed surface craft in the littoral to counter enemy antiaccess strategies. Subsequently, NWDC leased a high-speed vessel for FBE-I and FBE-J,[12] to experiment with the HSV in conjunction with various payloads to determine its merit.

FBEs are used not only to investigate whether new concepts and technologies have utility, but also to find out whether a concept makes sense in its formulation. The focus of an experiment may be doctrine and TTPs coevolved in association with a new technology. The results of FBEs can be used to accelerate the delivery of new DOTMLPF to the fleet. Alternatively, results can be used to shape more experimentation, to drive additional research, or to terminate efforts that do not warrant future investigation or investment.

FBEs and LOEs have different schedules, complexity, and resource requirements. LOEs are used to examine a single (or at most a few) well-defined projects or concepts in situations in which a broad range of operational parameters can be

[12]Fleet battle experiments are named by the Navy's phonetic alphabet. A = Alpha, B = Bravo, C = Charlie, D = Delta, E = Echo, F = Foxtrot, G = Golf, H = Hotel, I = India, and J = Juliet.

examined without the constraints of time and resources that are inherent in large-scale FBEs. In recent years, FBEs have become progressively more complex and have incorporated progressively more tests and experiments within limited periods of platform and asset availability. As a consequence, it can be argued that the experiments undertaken during FBEs are inherently more incomplete than those carried out during LOEs. For example, a complete examination of the operational parameters and attributes of the HSV could not possibly have been carried out during the few-week period associated with an FBE. An LOE dedicated to an examination of the attributes of the HSV was used in addition to the vessel's participation in FBEs. As a result, the Navy has a much broader understanding of HSVs than it would have had if the HSV had been examined only during the course of an FBE.

FBEs typically require 12 to 18 months. Platform and equipment availability must be planned in conjunction with personnel training cycles and operator availability. Surrogates, computer simulations, or models must be used when needed equipment is not available to support the experiment. These must be tested and verified as faithful representations of the system or capability being simulated or explored.

The FBE process involves many steps. They include the determination of objectives, concept definition, venue identification, selection of initiatives, technology selection, detailed planning, supporting events such as war games, simulations, and then the refinement of experiment planning, detailed preparation, execution, and evaluation.

For FBE-A through FBE-J, the process began with the solicitation of inputs from regional combatant commanders (then referred to as Commanders in Chief (CINCs)) for a numbered fleet sponsoring. The choice of sponsorship is synchronized with scheduled exercises owing to the need for live forces. Once selected, a sponsoring numbered fleet commander advises NWDC of warfare priorities and geopolitical and operational issues for the experiment. In response, NWDC recommends additional areas for consideration. Suggestions are also collected from the fleet, OPNAV, SYSCOMS, combatant commanders, ONR, the Navy laboratories, the Defense Advanced Research Projects Agency (DARPA), and industry.

Each FBE has a budget determined by its scope. Typically, FBE costs are between \$3 million and \$5 million[13] (these are costs beyond those of fleet operations and prototype system development for the FBE). In contrast, the Navy component for large joint experiments is typically about \$16 million. Funding for an FBE only pays for personnel support, supporting communications architectures, and technology.

---

[13]CAPT Patrick Denny, USN, Director, Maritime Battle Center, "Navy Experimentation," presentation to the committee on July 9, 2002.

Funding for an FBE is provided by NWDC; its funding is supplemented by various sponsors within the Navy, such as ONR, whose primary responsibility is the management of the Navy's science and technology (S&T) program, and by organizations external to the Navy such as DARPA. ONR provides some support for NWDC's planning of FBEs and frequently serves as a sponsor for individual projects and concepts that are evaluated in an FBE. From ONR's perspective, NWDC is a claimant for available funds. Although ONR intends to provide from $20 million to $40 million per year for experimentation, in the past such funds have been used for other, higher-priority efforts. Such preemption has occurred with other sponsors as well.

As noted, many organizations participate in FBEs. NWDC designs the overall experiment, the concept, and the scenarios around which the concept is to be tested. Numbered fleet command personnel arrange for support services. Personnel from host ships are assigned to train on test equipment before an experiment is initiated and to assist with shipboard installation. SYSCOM representatives examine all temporary installations for safety. New equipment and systems are tested, integrated, and maintained once the experiments have commenced. Since there is no defined logistic support system available for experimental equipment undergoing testing, contractors normally provide support for their equipment with senior engineers and technicians. Individual project offices that are sponsoring systems or equipment normally provide their own teams of observers and analysts. The personnel from many organizations—military and civilian, government and contractor—monitor and observe events during the experiment, collect data, conduct analyses and evaluation, and prepare the lessons-learned and after-action briefings. For instance, when their programs or responsibilities are involved, ONR observers attend FBEs, review the final reports, and recommend changes and adjustments to ongoing R&D efforts based on the results. An indirect benefit from the level and nature of contractor involvement is the influence on industry independent research and development (IR&D).

Finally, since all systems and equipment being tested in an FBE must have an intended transition recipient (e.g., a SYSCOM or a program executive office), representatives of such organizations are also present. In short, many personnel contribute to the planning, preparation, execution, and evaluation of an FBE or an LOE.

While final versions of reports require long preparation times, lessons learned are prepared shortly after the FBE's completion and are briefed to fleet operators, senior personnel of the ships and aircraft involved, and senior leadership in OPNAV. After each FBE, the sponsoring fleet sends a message to all major naval commands that summarizes "quick-look" results. ONR, all major SYSCOM organizations, and program executive offices are on the distribution list, as are the N7 and N8, since these organizations have the ultimate responsibility for sponsoring, budgeting for, and developing new capabilities.

SYNOPSIS OF RESULTS TO DATE FROM FLEET BATTLE EXPERIMENTS ALPHA THROUGH INDIA

In FBE-A through FEB-J, the U.S. Navy experimented with concepts of operations and with new TTPs using networked systems for theater ballistic missile defense (TBMD), naval fire power, and defense against asymmetric threats. Each FBE had between three and eight major objectives, and each major objective had anywhere from one to nine subobjectives.

Table 3.1 summarizes the major objectives and findings of FBE-A through FBE-I. A complete analysis of the results of FBE-J was not available to the committee at the time of its study.

Table 3.1 is necessarily telegraphic, merely summarizing results. A review of the results reveals that, while all objectives might not have been achieved, a significant number of objectives were indeed realized. The reader should ask, of course, whether the lessons learned were trivial or profound, whether they could have been learned more easily and less expensively in other ways (such as through LOEs—a point discussed above with respect to the investigation of the HSV). By and large the committee concluded that the FBEs have in fact been valuable—and possibly invaluable because of their unique ability to focus the attention and excite the imagination of important senior officers with a warfighting perspective.

One of the important outcomes of the Navy's FBE campaign has been the determination as to why certain objectives were *not* achieved, which allows for an iterative process of improvement. The principal successes of FBEs have been as follows:

- Demonstrations of the feasibility of new operational concepts using surrogate or prototype systems or existing systems,
- The adoption by fleet forces of new TTPs, and
- The development of new doctrine for fleet operations.

A Recent Example of New Concept Development

FBEs have supported the evolution of decision-support concepts and tools intended to increase the speed of command[14] and enable the collaboration of command echelon decision makers. The Knowledge-Web (K-Web) addressed in FBE-J is one such example. The K-Web involves the application of knowledge-management practices to warfighting, creating a concept of operations in which value-added information (i.e., "knowledge") is created and published on the command intranet in real time rather than being coupled to daily briefing cycles.

[14]Speed of command can be defined as the rapidity with which decisions are made by all the ships involved in making command decisions, the decisions are formulated as executable orders, and the orders are communicated to those responsible for their execution.

TABLE 3.1 Summary of Major Objectives and Findings of Fleet Battle Experiments Alpha Through India (FBE-A Through FBE-I)

FBE Name, Year Conducted	Objective	Findings
FBE-A 1997	1. Determine whether Commander, Marine Air/Ground Task Force could make tactical decisions based on Common Tactical Picture (CTP) seen on the USS *Coronado*.	1. Tactical decisions based on CTP seen on the USS *Coronado* were not possible because one-way data link limited distribution of CTP.
	2. Demonstrate collaborative planning between Third Fleet and USAF, Special Purpose Marine Air/Ground Task Force (SPMAGTF), Area Air and Missile Defense Command, USMC Hawk, and the U.S. Space Command (SPACECOM); evaluate automated support and decision tools for deliberate and reactive planning.	2. Collaborative planning was performed among all entities except the ashore SPMAGTF. No collaborative planning was demonstrated for Naval Fire Support. Collaborative planning was demonstrated for theater ballistic missile defense (TBMD), as it was for Joint Task Force Exercise-97.
	3. Examine Naval Surface Fire Support operational concepts utilizing a simulated arsenal ship, new command and control architectures, and advanced ordnance.	3. The command and control relationships between the Joint Force Air Component Commander (JFACC) and the Naval Fire Cell were a source of contention and discussion. JFACC could easily integrate employment of arsenal ship in Rapid Strike scenario and successful results were achieved. Marine operators handled a peak of 2.5 targets per minute, error free.
FBE-B 1997	1. Examine and test concepts of operations for controlling all fire missions through defined threads using advanced sensors and technology.	1. The Naval Simulation System (NSS) simulation was used to provide a means of testing the Ring of Fire concept with a flow of target nominations similar to that in a wartime situation. The distributed C4ISR architecture using the Tactical Real-Time Targeting System was substantially more efficient at nominating targets for engagement than is the current, centralized architecture.

continues

TABLE 3.1 Continued

FBE Name, Year Conducted	Objective	Findings
FBE-B 1997	2. Examine and test the use and value of the Common Operating Picture/Common Tactical Picture (COP/CTP).	2. The feasibility of using the Joint Maritime Command Information System to display the COP/CTP at all Navy command and control (C2) nodes was demonstrated.
	3. Examine and test new procedures for deconfliction of multiple platform launches and existing flight plans.	3. NSS identified actionable targets and nominated them to Land Attack Warfare System (LAWS) for allocation to surface fire and close air support engagements; NSS also simulated target nominations to LAWS from four Forward Observer Air Controller Systems operating near the battlefront on the islands off the California coast.
FBE-C 1998	Explore alternative tactics, techniques, and procedures in executing the Ring of Fire concept for Joint Fires support in a littoral environment.	Fires Cell teams were exercised while testing alternative functional arrangements and procedures within the Fires Cell; increasing improvement in performance (as measured by time needed to service targets) validated the procedures and techniques tested.
FBE-D 1998	Examine (1) improved detection and targeting of Maritime Special Operations Forces through integrated use of air, surface, and subsurface Combined Forces under the Ring of Fire concept; (2) improved theater situational awareness and execution through coordination and application of a joint tactical picture; and (3) seamless sharing of a CTP between components for the Counter Special Operations Forces mission.	Bandwidth was limited and heavily used; new systems need to use it conservatively. The effect of added network load on legacy systems needs to be assessed. File transfers used in collaborative planning represented a trade-off between communication delays and bandwidth availability; all observed delays appeared to be acceptable during FBE-D.

TABLE 3.1 Continued

FBE Name, Year Conducted	Objective	Findings
FBE-E 1999	1. Explore Full Dimensional Protection.	1. Utilization of the unmanned aerial vehicle (UAV) for detection, identification, and tracking had remarkable value; a combat swimmer could be detected by the UAV; high-quality imagery of mobile targets was almost continuously available to the Harbor Defense Commander, Full Dimension Protection Cell, and others.
	2. Explore response to asymmetric threat.	2. Embarkation of the Mobile Inshore Undersea Warfare van extended organic and inorganic sensor range and allowed it to be used in the littoral zone of interest without having to establish a secure rear area for Mobile Inshore Undersea Warfare Unit protection.
	3. Explore network-centric antisubmarine warfare (NCASW) with collaborative multisensor planning.	3. NCASW increased force situational awareness through distributed advance search plans. Reliable networked communications are essential for distributed collaborative planning in NCASW. Common tactical decision aids enhance the update of situational awareness required for NCASW.
	4. Explore Theater Air and Missile Defense.	4. Changes in tactics are needed to compensate for arc and range of fire. Deconfliction requires further investigation. Improved identification methods are needed to prevent fratricide of high-value-asset defenders and to take into account possible collateral damage both over water and ashore.
	5. Explore precision engagement.	5. Weapons currently used for naval surface precision fires were not found to be useful against targets in urban canyons.

continues

TABLE 3.1 Continued

FBE Name, Year Conducted	Objective	Findings
FBE-F 1999	1. Demonstrate ability to counter undersea threat and maintain sea line of communication into Persian Gulf.	1. Current C4I does not support requirements for in-stride mine clearance operations. The colocation of the Sea Combat Commander and the Mine Warfare Commander with the Joint Force Maritime Component Commander (JFMCC) was effective. Full exploitation of shared sensors and environmental data was not realized.
	2. Demonstrate ability to protect Mine Countermeasures (MCM) forces by use of integrated joint forces assets.	2. Army attack aviation (e.g., Apaches) appeared uniquely fitted to this role of engaging surface threats. Issues involve airborne C2 and the ability of Apaches to discriminate between friendly and hostile targets require resolution.
	3. Demonstrate ability to protect MCM forces by application of disruptive, neutralization, and suppressive fires to the shore-based threats.	3. Joint Fires Element (JFE) of the Commander Joint Task Force (CJTF) staff demonstrated the ability to control both deliberate and tactically responsive fires. JFE appeared particularly fitted to control of Navy fires during initial penetration when the preponderance of fires was maritime-based.
	4. Examine implications of effects-based operations (EBO).	4. The integration of EBO was not effective owing to a lack of common understanding of EBO and a common language, and inadequate definition of requirements for detailed, continuous commander's guidance.
	5. Examine Joint Task Force nuclear, chemical, and biological defense capabilities.	5. The land-based CJTF was particularly vulnerable to nuclear, biological, or chemical (NBC) attack delivered by either theater weapons (i.e., theater ballistic missile) or by terrorists; the experiment illuminated the extent of the burden of conducting operations in an NBC environment.

TABLE 3.1 Continued

FBE Name, Year Conducted	Objective	Findings
FBE-G 2000	1. Examine capability of Composite Warfare Commander to provide seamless transition from tactical-level defense of battle group to theater defense of civilian population centers.	1. Navy assets assigned to theater ballistic missile defense missions produce C2 ambiguity when they report to theater commanders for TBMD and tactical commanders for other mission areas.
	2. Examine use of Special Operations Forces (SOF) in time-critical target (TCT).	2. SOF demonstrated (1) the use of unattended ground sensors and an airborne Synthetic Aperture Radar/ Multispectral Thermal Imagery sensor, (2) capability to perform intelligence preparation of the battlefield through use of unattended ground sensors, and (3) a strike-coordinating role by an SOF cell deployed onboard a submarine.
	3. Demonstrate use of combined strategic, operational, and tactical-level sensors in a single network to improve early warning missile defense time lines (sources for ballistic missile (BM) tracks included Tactical Data Dissemination System, Joint Tactical Ground Station, Tactical Exploitation of National Capabilities (TENCAP), airborne laser, Aegis, Theatre High Altitude Area Defense, and Patriot).	3. Cueing data could be used to dedicate radar resources in affected sensors and to confirm or deny existence of BM tracks; use of multiple sensors in a single-sensor network significantly decreased time to validate BM tracks and improve early warning time lines; Navy's TENCAP systems and processes decreased validation times for BM tracks by feeding Joint Tactical Air-to-ground System raw infrared data to an Aegis destroyer.
	4. Examine the tactical employment of a network-centric warfare engagement network created using a combination of command and control options in the digital fires network.	4. The experimental C2 network demonstrated the commander's requirement to impose varying degrees of control, depending on the tactical situation.
	5. Examine application of antisubmarine warfare (ASW) search methods to TCT.	5. ASW search techniques could be used.

continues

TABLE 3.1 Continued

FBE Name, Year Conducted	Objective	Findings
FBE-G 2000	6. Build and maintain an overland Common Operating Picture, identified by Commander Sixth Fleet as a key shortfall during Operation Allied Force.	6. This objective was not achieved.
	7. Examine limitations of a distributed mensuration network to develop aim points for decentralized engagement.	7. Mensuration was achieved with average time of 9.5 min; despite high-bandwidth, low time-latency networks, the speed of execution (sensor to shooter) could not adequately engage targets with dwell times of less than 30 min.
FBE-H 2000	1. Examine a deliberate targeting and planning process based on collaboration between the warfare commanders to produce a Maritime Tasking Order (MTO) that provides guidance, apportionment/allocation instructions, and deliberate assignment of targets.	1. A collaborative planning process to prioritize, deconflict, and synchronize future maritime missions within an acceptable planning cycle was demonstrated. FBE-H indicated a need for planning and collaboration tools resident with each warfare commander and the JFMCC staff. The relationships between the Joint Force Air Component Commander and JFMCC and the relationships between the air tasking order and MTO were not examined.
	2. Examine the application of the Digital Fires Network (DFN) to the time-sensitive-target problem. DFN was a synthesis of C2, intelligence, surveillance, and reconnaissance (ISR), and fire support planning and execution tools, with LAWS, Global Command and Control System (GCCS) ISR Capability, and Precision Targeting Workshop (PTW) being the core systems.	2. (a) A COP proved elusive—in varying degrees LAWS, GISRC, Global Command and Control System-Maritime (GCCS-M), and numerous warfare-area-specific tools contain information that is required for a thorough understanding of the battlespace, and during FBE-H the information required for situational awareness was present but could not be easily aggregated or shared between systems.

TABLE 3.1 Continued

FBE Name, Year Conducted	Objective	Findings
FBE-H 2000		2. (b) The value of conducting the experimentation with joint and allied forces was demonstrated (HMS *Cardiff* was integrated into the Digital Fires Network via a LAWS terminal); DFN could be used by Army and Marine Corps personnel to plan for and execute Tactical Tomahawk Land Attack Missile missions (weapons were allocated, assigned to targets, launched to loiter points, and retargeted to support the maneuver of Army and Marine forces).
	3. Test the ability of the Mine Warfare Commander (MIWC) and Mine Countermeasures Commander (MCMC) to execute the mine warfare mission in support of the aircraft carrier battle group operations.	3. With some deficiencies, the concept for using organic MCM systems hosted on platforms with other warfare missions worked well in clearing the mined SLOC and approaches to the amphibious assembly area. As currently configured and tasked, a submarine development squadron (DESRON) staff is incapable of performing additional MIWC duties without a significant increase in manning and mine warfare expertise.
	4. Demonstrate the operation of GCCS-M using a Transmission Control Protocol/Internet Protocol (TCP/IP) network architecture equipped with the COP Synch Tool, and a database replicator, to provide low-latency replication of the GCCS picture across the battle force.	4. A low-latency GCCS-M data distribution throughout the COP was achieved but did not translate into a common understanding of the battlespace for commanders and planners; as configured, GCCS-M required dedicated management to ensure that information was complete, timely, and accurate.
	5. Demonstrate the effective employment of nonlethal technology during simulated Maritime Interception Operations (MIO).	5. Nonlethal technology was effectively employed during simulated MIO; nonlethal means proved effective in gaining compliance of a live crew during a permissive but noncompliant experimental MIO scenario.

continues

TABLE 3.1 Continued

FBE Name, Year Conducted	Objective	Findings
FBE-H 2000	6. To collect data in three major ASW areas: (1) sensor fusion, (2) use of remote and autonomous sensors, and (3) techniques to collect meteorological data in denied areas.	6. Remote/autonomous sensor fields in the form of simulated Long Endurance Low Frequency Active Sonar (LELFAS) buoys showed promise when used to shape the battlespace and as a force multiplier by freeing other ASW forces; LELFAS was successfully employed to monitor high-value unit operating areas and to drive adversary submarines into areas where their weapons and sensors were less capable.
FBE-I 2001	1. Assess the value of the Joint Medical Operations-Telemedicine (JMO-T), an ACTD whose purpose is to facilitate administrative and operational aspects of medical care.	1. System users believed that the JMO-T system shows tremendous promise for field use; in its present form it is not robust enough and self-compatible enough within its entirety to be close to field introduction.
	2. Examine a counterforce concept that uses a tiered sensor system including live Special Operations Forces.	2. The concept of a tiered sensor system including live Special Operations Forces was found to be sound; execution of this initiative brought to light technical and logistical limitations that, when combined, were quite challenging; expectations for performance, though perhaps unreasonably high, were not met.
	3. Explore two initiatives for assured access ASW, specifically, network-centric coordinated antisubmarine warfare (NCCASW) and submarine-launched UAVs (SLUAVs).	3. NCCASW provided a capability for both manned and unmanned sensors to increase the battle group's situational awareness of ASW assets and of the threat. The SLUAV provided a distributed ISR capability in the battlespace, particularly for post-strike battle damage assessments and coastal/port surveillance and for time-sensitive, quick-reaction launches.

TABLE 3.1 Continued

FBE Name, Year Conducted	Objective	Findings
FBE-I 2001	4. Test six measures of performance (MOPs) intended to assess whether the Information Knowledge Advantage (IKA) features helped improve network-centric warfare.	4. The MOPs were sufficiently complete to cover all of the pieces of the network-centric architecture, including new communication architectures, new application delivery options and underlying databases, and new methods of discovering and retrieving data and human interfaces. Assessment of the MOPs led to generally positive conclusions regarding the ability to transform information into knowledge and an eventual combat advantage.
	5. Demonstrate the possibility of transforming network-centric warfare to standard naval practice.	5. The conditions for the transformation of network-centric warfare to standard naval practice were established by the combination of a tactically and technically proficient Joint Task Force and Component staff, the Digital Fires Network, an extended battlespace, and a robust experimental concept of operations.
	6. Demonstrate the integration of the supporting concepts of Joint Fires to include the Tactical Exploitation System-Navy (TES-N), as a component of the Naval Fires Network.	6. (a) TES-N was shown to be able to operate in a relatively high tempo, warfighting environment networked in the FBE-I architecture that included C4ISR and weapons-target-pairing systems. TES-N contributed to a complete sensor-weapon capability for FBE-I that exploited strategic, operational, and tactical sensor products. (b) The primary systems resident in the FBE-I architecture that enhanced the capabilities of TES-N were the GCCS-M, GISRC, Ku band network, and LAWS.

Government off-the-shelf tools have been developed to take the work out of creating and maintaining networked content. They include a template-based authoring tool, Summary Maker, and a graphical drawing and annotation tool known as TACGRAPH that creates map-based information products—tools that allow information to be easily authored and disseminated in a manner consistent with a command's current business practices.

The K-Web concept was initially developed for the Global 2000 and Global 2001 War Games. Based on the utility of the K-Web and supporting tools, the Commander, Carrier Group Three asked that revised versions of the tools be deployed and placed aboard the USS *Carl Vinson* during its fall 2001 deployment. The SPAWAR Systems Center (San Diego), with the support of the Office of Naval Research, developed and installed a prototype system aboard the USS *Carl Vinson* in less than 5 months.

On September 11, 2001, the Commander, Carrier Group Three, assumed command of Task Force 50 in the North Arabian Gulf. As a result, the K-Web was battle-tested by Carrier Group Three during Operation Enduring Freedom. Rear Admiral Thomas Zelibor, as Commander of Carrier Group Three, found the K-Web to be a "powerful" tool in the conducting of Operation Enduring Freedom, and the entire region came to rely heavily on the products stored in the Carrier Group Three K-Web. Following the return of the carrier group from its deployment, SPAWAR, the Commander in Chief of the Pacific Fleet, the CFFC, as well as Carrier Group One and the Commander, Third Fleet Network-Centric Innovation Center have worked closely to migrate the K-Web tools to SPAWAR programs of record (specifically Global Command and Control Systems-Marine (GCCS-M)) and to transition the K-Web to additional battle groups. The K-Web is currently being integrated with the Collaboration at Sea program. The first release of the K-Web with instructions for its reproduction was used by the USS *Constellation* battle group during its fall 2002 deployment and is planned to support the upcoming deployments of the USS *Nimitz* and USS *Theodore Roosevelt* battle groups.

The Chief of Naval Operations has directed that the Navy become "web-enabled" and that it work toward maintaining "knowledge superiority." The latest thrust of this effort is known as "FORCEnet." The K-Web represents a significant first step in achieving these goals in that it defines and has demonstrated important progress toward a new concept of operations for warfighting. Because it is explicitly designed to support the distributed collaboration that is becoming core to modern military operations, the K-Web is directly relevant to the efforts of Task Force Web and FORCEnet, as well as to numerous command and control programs including GCCS-M and the Collaboration at Sea initiative.

Transition into Acquisitions

The process of tracking the fate of products and technologies that performed successfully in FBEs has proven difficult. The committee was unable to identify the transition of any such products or technologies directly into the Navy's acquisition process in the sense of their leading to new programs in the Navy's program of record. This is a significant shortfall. The failure derives from several factors, including the following:

- The requirements process that must be satisfied before an effort can become an acquisition program;
- The relative shortness of time since the completion of some experiments (e.g., FBE-J was completed in the summer of 2002, and final reports were not disseminated until early 2003); and
- The Navy's program of record, which represents the result of many compromises, trade-offs among many programs competing for scarce funding, and prioritization by the Navy's senior leadership.

However, FBE results have had some influence on acquisitions. For example, the results of the FBE-I and FBE-J explorations of the HSV-X1 have had an impact on the design concept of the littoral combat ship. In a sense, the route of entry of the results of FBEs seems to be either through evolutionary upgrades of existing systems or through modification of the design concept or implementation plans of ongoing programs of record.

The recent introduction to the fleet of the Naval Fires Network (NFN), first tested in FBE-A and later refined in FBE-I, has the potential for the most successful transition of the results of an FBE-tested concept. However, to date NFN has not become a formal program of record.

NFN is a network-centric warfare system that provides real-time intelligence correlation, sensor control, target generation, mission-planning, and battle damage assessment capabilities. It allows ships and aircraft in a carrier battle group (CVBG), amphibious ready group (ARG), or expeditionary strike group (ESG) to share near-real-time and real-time intelligence and targeting information not only with one other, but also with Army and Air Force units in a joint or coalition task force.

The successful test of NFN in FBE-I and its subsequent introduction into the fleet as an interim prototype on two CVBGs demonstrate that FBEs may become a vehicle for the rapid introduction of new capabilities into the fleet.

Future Plans for Fleet Battle Experiments (FBE-Kilo and Beyond)

FBE-A through FBE-J were developed under the hierarchy of planning called Concepts-Based Experimentation. Concepts were proposed at a high level by

NWDC and built momentum that would extend throughout the experiment planning cycle. The maturation and development of a concept could extend up to the final planning conference for the experiment.

Initiatives, derived from concepts, were also kept at a high level. An initiative might be stated in terms such as, "Examine the use of Digital Fires Network on Time Sensitive Targets." Experimental objectives then followed from an initiative description. In some cases, new initiatives and objectives were also introduced up to the final planning conference.

Experimental questions and hypotheses were usually derived from simulations, analytical studies, and war games that then provided the details necessary to conduct an experimental design, data collection, and analysis planning. The continuous changes in concepts and objectives place pressure on the design of the experiment.

An adjustment to this process was taking place at NWDC with the planning for Fleet Battle Experiment-Kilo (FBE-K); see Figure 3.1. Concepts were developed by Warfare Innovation Development Teams (WIDTs), each of which was assigned a high-level concept that NWDC was interested in exploring at the time—Navy Fires and the Information and Knowledge Advantage concept.[15] When concepts flowed, doctrine was associated, and a set of related objectives was developed for an FBE. Concepts and associated objectives were produced at the front end of the experimentation process in order to be more stable throughout that process. Each WIDT was responsible for a set of activities, which consists of war games, meetings, LOEs, exercises, real-world experiences, and so on. FBEs may or may not be the culminating event.

The committee believes that these changes produce several positive consequences. They improve stability for good experiment design and place emphasis on selecting the right kind of activity to address the problem at hand. Nonetheless, there are concerns with this process:

- The process by which a concept becomes worthy of further exploration by a WIDT is not clear. The pool of nominations is deep and comes from many sources, as discussed earlier. Considerations of sponsors and funding, technical maturity, and feasibility are among the many factors that are weighed in the selection process, which is dependent on human judgment.
- Although more time may be available for better experiment design and planning as a result of the changes made, the process must be stable enough to produce better results.

[15]The committee understands that NWDC has recently changed this process somewhat. There are still five WIDTs, now called Sea Strike, Sea Basing, Sea Shield, Information and Warfare Advantage, and Combating Terrorism/Force Protection. The committee believes its comments are applicable to this process even if organized differently.

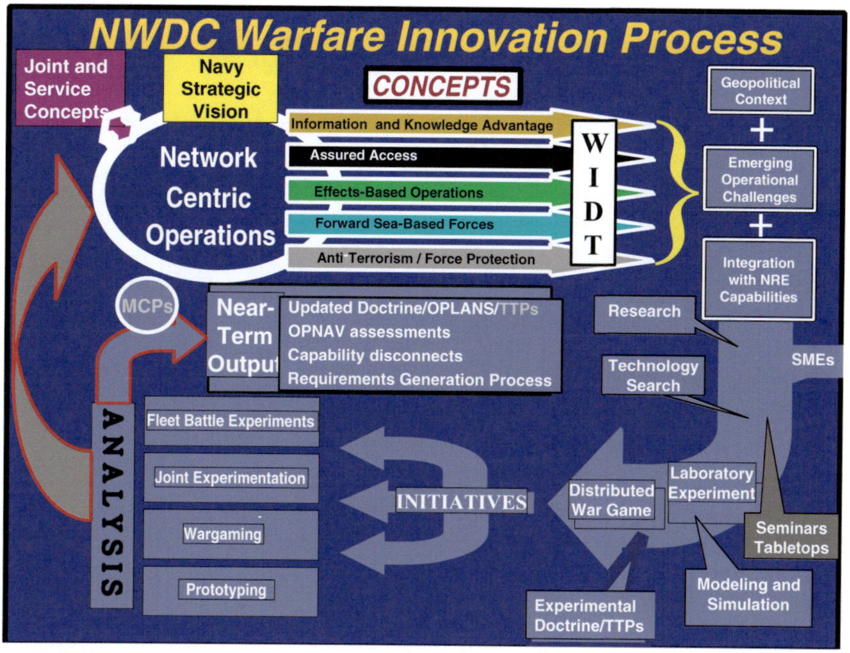

FIGURE 3.1 Warfare innovation process of the Navy Warfare Development Command.

NOTE: WIDT = Warfare Innovation Development Team; MCP = Mission Capabilities Package; OPLANS = operational plans; TTPs = tactics, techniques, and procedures; NRE = Naval Research Establishment; SME = subject matter expert.

- It is also not clear how feedback and knowledge gained from other events and venues will contribute to the definition of new concepts and experimentation objectives and result in coherent experimentation campaigns. While the importance of campaigns was clearly acknowledged, no example of how the process results in experimentation campaigns was forthcoming.
- Participation in the WIDTs was still primarily an internal function of NWDC, without much participation from outside stakeholders.
- Analysis and objectives were still emphasized at the end of the experimentation process, rather than at the beginning.
- Cross-concept analysis was further fractured, as there was not a clear relationship between the learning from one WIDT and that from another, or between experiments developed in different WIDTs.

FUTURE EXPERIMENTATION FOR NAVAL TRANSFORMATION

Today's security environment is characterized by uncertainty, surprise, and conflict. *Joint Vision 2020*, the joint defense community's principal template for future capabilities, calls on the U.S. Armed Services to prepare for an uncertain future by creating forces that are faster, more precise, and more lethal than today's. [16] The report sees innovation in technology, organizations, and concepts as vital to the process of creating new operational capabilities.

Similarly, the Quadrennial Defense Review of 2001 calls on the military to transform the way it operates. The review identifies six broad operational goals for transformation, including the protection of critical bases, assurance of information systems, projection of forces in the face of antiaccess threats, persistent surveillance and tracking, rapid precision engagement, enhanced space capabilities, and interoperable, joint command, control, communications, computers, intelligence, surveillance, and reconnaissance (C4ISR).[17] The implication is more than modernization. It includes the creation of new fighting concepts and packages of military capability, potentially involving every element of DOTMLPF.

As indicated in Chapter 1, the CNO's capstone concept, Sea Power 21, includes Sea Strike, Sea Shield, and Sea Basing as elements incorporating the key offensive, defensive, and support elements, respectively, of network-centric warfare. Experimentation is a key enabler under the heading of Sea Trial. The *Naval Transformation Roadmap* lays out a blueprint to guide Navy and Marine Corps efforts in support of those goals.[18]

- Sea Strike is a concept for projecting offensive power from the sea in support of joint objectives. Its transformational capabilities include persistent intelligence, surveillance, and reconnaissance (ISR); time-sensitive strikes; information operations; and ship-to-objective maneuver.
- Sea Shield exploits naval control of the seas and forward-deployed defensive capabilities to defeat antiaccess threats, enabling joint forces to project and sustain power. Its transformational capabilities include theater air and missile defense (TAMD), littoral sea control, and homeland defense.
- Sea Basing is intended to provide sustainable global power projection from the high seas at the operational level of war. Its transformational capabilities

---

[16]GEN Henry H. Shelton, USA, Chairman of the Joint Chiefs of Staff. 2000. *Joint Vision 2020*, The Pentagon, Washington, D.C., June, p. 13. Available online at <http://www.dtic.mil/jointvision/jv2020.doc>. Accessed October 7, 2003.

[17]Donald H. Rumsfeld, Secretary of Defense. 2001. *Quadrennial Defense Review Report*, Washington, D.C., September 30, Ch. V.

[18]Secretary of the Navy Gordon England, Chief of Naval Operations Vern Clark, and Commandant of the Marine Corps James L. Jones. 2002. *Naval Transformation Roadmap: Power and Access . . . From the Sea*, Washington, D.C.

include accelerated deployment and employment times for power projection assets and enhanced seaborne positioning of joint assets.

- Sea Trial is the Navy's nomenclature for its processes for innovation.[19] Through Sea Trial, the Navy plans to rely extensively on experimentation, with the goal of rapidly delivering emergent technology, doctrine, and capability to the fleet. The Sea Trial process is fleet-led under the CFFC, and NWDC is designated as the project coordinator for the entire Sea Trial process:

> NWDC will work closely with the Program Executive Offices, Systems Commands and designated units to integrate these options into practice, developing and testing the capabilities in fleet battle experiments and joint exercises, culminating in operational deployments. . . . Thus the fleet commander, NWDC, and designated fleet units form an interactive team that carries innovation from the laboratory to deployment with focus, speed and efficiency.[20]

FORCEnet

As a fully implemented physical entity, FORCEnet does not currently exist. Conceptually it will be the network and its associated architecture, interface standards, and protocols, which will integrate warfighters, weapons, sensors, databases, and decision aids into a comprehensive warfighting maritime system.

The design of such a system will require many complex trade-offs. Computer simulations, analytic studies, and at-sea tests will be required to allow appropriate and optimal choices to be made. Ultimately, many of these choices will be made as a result of operational experience derived from many future LOEs and FBEs, indeed from a coherent set of experimentation campaigns.

The organizations within the Navy (Navy Network Warfare Command as the FORCEnet Type Commander and N61/N704 as the Resource and Warfare Sponsor/Joint Interoperability) that have responsibility for transforming FORCEnet into an operational capability are committed to such rigorous and extensive campaigns of experimentation (supported by computer simulation and analytic studies), which will result in a rapid convergence on FORCEnet needs and in the rapid acquisition of those components that must be procured. N61 has programmed a number of LOEs beginning in March 2003 that will continue with frequent interim evaluations through February 2004. If the proposed experimentation is successful, a prototype development will be provided to a battle group and an amphibious ready group by the fourth quarter of FY 2004 for operational evaluation.

---

[19]ADM Vern Clark, USN, Chief of Naval Operations. 2002. "Sea Power 21: Projecting Decisive Joint Capabilities," *U.S. Naval Institute Proceedings*, Vol. 128, No. 10, October, pp. 32-41.

[20]Secretary of the Navy Gordon England, Chief of Naval Operations Vern Clark, and Commandant of the Marine Corps James L. Jones. 2002. *Naval Transformation Roadmap: Power and Access . . . From the Sea,* Washington, D.C., p. 34.

U.S. MARINE CORPS

Past and Present Experimentation

Historically, the Marine Corps has used experimentation to develop new capabilities or to bring about changes in equipment, doctrine, tactics, training, and procedures. Early efforts led to the use of aircraft for close air support for ground troops during the 1920s, the development of amphibious warfare doctrine during the 1930s, the use of helicopters in combat during the 1950s, the development and use of vertical-capable jet aircraft during the 1970s, the development of very short takeoff and landing rotorcraft during the 1980s, and the building of Fleet Antiterrorism Security Teams and the Chemical/Biological Incident Response Force of today.

During the 1990s, the Marine Corps renewed its efforts in experimentation with General Charles Krulak's establishment of the Commandant's Warfighting Laboratory (CWL). The principal objective of the laboratory was to experiment with advanced warfighting capabilities that were generated from the Concept Based Combat Development System. The purpose of the experimentation program was to explore responses to requirements deficiencies; examples of such responses are a lightweight mobile fire support system or an intraplatoon/squad communications system to enhance advanced warfighting capabilities.

Marine Corps experimentation addressed personnel, organization, doctrine, tactics, training, procedures, and equipment—all components of DOTMLPF. However, resources for the experimentation program were limited, and the scope of the program fluctuated depending on financial, personnel, and operational tempo conditions.

The Marine Corps experimentation program of the 1990s was called Sea Dragon; a 5-year campaign plan was developed. This plan began with the standup of the CWL and the establishment of a Special Purpose Marine Air/Ground Task Force Command Element (SPMAGTF(X) CE). The CWL had direct responsibility for the development and execution of the Marine Corps Experimentation Warfighting Program. The SPMAGTF(X) CE Headquarters worked directly for the Commander, CWL (since 1995, Commanding General, Marine Corps Warfighting Laboratory (MCWL)) and was the field execution agency headquarters for the MCWL. The 5-year plan laid out in broad terms a series of advanced warfighting experiments tied directly to the deficiencies noted in the capabilities required to realize the advanced warfighting concepts of the Marine Corps.

Methodology

The major intellectual factor influencing the early stages of the Marine Corps experimentation program was the Marine Corps Combat Development System (CDS), a concept-based requirement system. Through this process, Marines analyze the range of threats anticipated in the future security environment, identify potential challenges, and then determine the warfighting requirements needed to address those challenges effectively. At the core of this process are concepts—which are formal documents that articulate the Marine vision for future warfighting. They look forward in time—beyond the concerns of today's programming and budgeting—and provide the spark that starts a focused process of proposal, debate, and experimentation. Marines use this participatory dialogue as the means of shaping the initial concepts, ultimately molding them into requirements that will provide the warfighting solutions needed. The hierarchy and relationship of Marine Corps concepts is shown in Figure 3.2.

Experimentation venues used by the Marine Corps included major advanced warfighting experiments (AWEs), preceded by warfighter discussions, focused discussions with subject-matter experts, seminars, symposiums, simulations, constructive simulations, limited technical assessments (LTAs), and LOEs—a spectrum of experimentation activities.

The Sea Dragon campaign was founded on the basic infantry warfighter and the basic warfighting unit—the Marine rifle squad. The campaign plan was

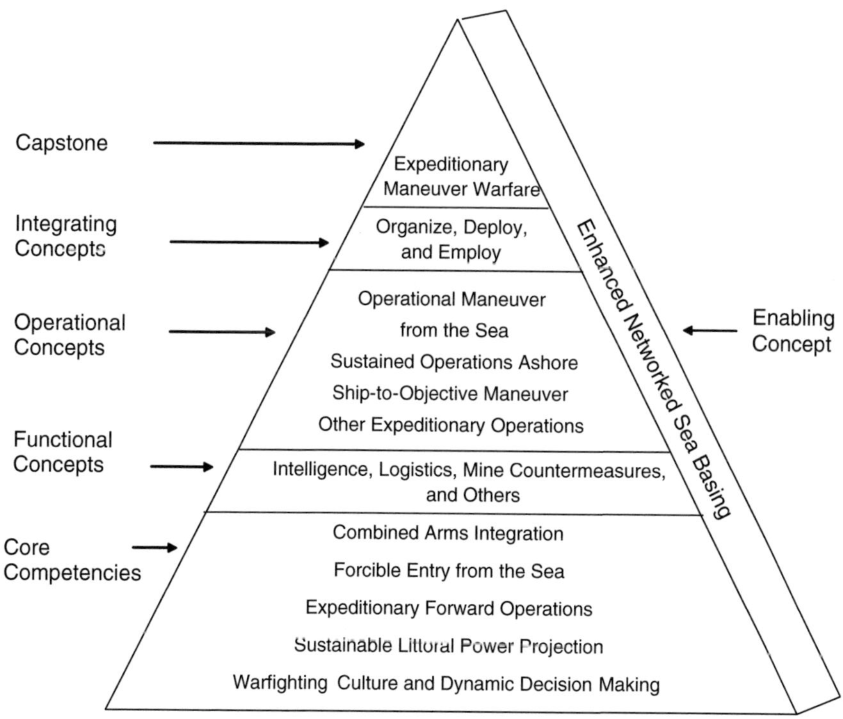

FIGURE 3.2 The hierarchy and relationship of Marine Corps capstone concepts.

designed first to improve the individual Marine's warfighting capabilities through experimentation with the introduction of new equipment, tactics, procedures, organizations, training, and doctrine. After this part of the campaign was well under way, the same basic process of experimentation was repeated with other Marine groups (fire teams, squads, combat patrols, and so on) to determine each unit's enhanced warfighting capabilities.

Sea Dragon included three major advanced warfighting experiments: Hunter Warrior (HW), which focused on individuals and combat patrols operating in desert environments; Urban Warrior (UW), which focused primarily on individual and platoon-size operations in an urban environment; and Capable Warrior (CW), which focused on individual and company-size operations at Camp Pendleton, California. Each series required a cycle of somewhat more than 3 years. The major advanced warfighting experiments were conducted about every 2 years, and overlaps occurred in experiments during the same period in time. Also, some of the early preparatory work (such as focused discussions and simulations) was repeated for each of the three stages (planning, execution, and analysis). Graphically the campaign cycle might look something like Figure 3.3.

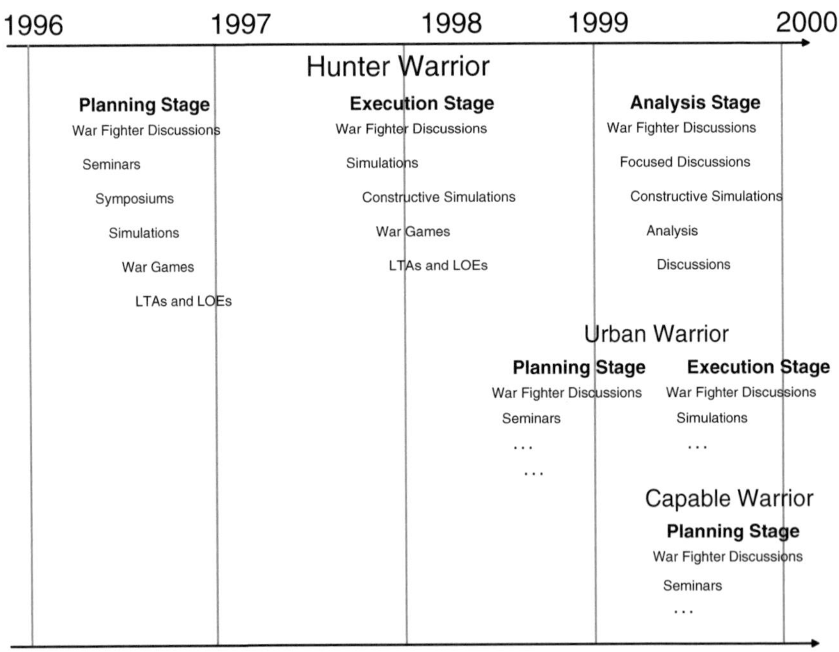

FIGURE 3.3 Marine Corps warfighting experiment campaign cycle.

In these campaigns, and even today, the Marine Corps makes extensive use of LTAs and LOEs. LTAs are focused on the technical performance of a piece of equipment; LOEs, on the utility of experimental TTPs in the context of a tactical scenario or on the utility of experimental technology in a tactical scenario. LTAs and LOEs are used to select down to "best of breed" for experimental TTPs, equipment, or systems.

Several LOEs and LTAs were conducted as interim spiraling experiments that were critical to each AWE in the campaigns. The LOEs served as affordable checks on progress or reality for specific concepts and for TTPs before their inclusion in the concepts, tactics, and procedures used in the larger field event. The main success of these large events depended, in turn, on the success of several interacting variables. War games were also conducted to provide a reality check for the specific experiments as well as to assist in training the forces.

The application of this strategy of using LOEs can be exemplified by experimentation with various nonlethal weapons during the Urban Warrior campaign. The Marine Corps was interested in the use of nonlethal weapons during military operations in urban terrain (MOUT) operations because of the large number of noncombatants normally found in urban areas. An LOE called the dazzler provided a simulated microwave/laser and noise capability (about 100 decibels) that incapacitated the target area. Marines and actors played the role of the targeted enemy, and 20 actual members of the media functioned as observers. The LOE was very realistic in its execution, and although there were no constructed deaths or actual injuries, the media had very mixed opinions on the type of stories they would write after viewing the dramatic effects and constructed agony of the target force. This LOE was very instructive with respect to the types and uses of nonlethal weapons, especially with noncombatants as part of the included target population. The results were applied in later LOEs and in the Urban Warrior AWE.

A second example illustrates the importance of LOEs as preparation for larger field events. RISTA (reconnaissance, intelligence, surveillance, targeting, acquisition) is a potential asset that would cover the battlefield. Part of its surveillance capability was provided by UAVs and unmanned ground vehicles (UGVs). Surveillance and reconnaissance ground equipment (SARGE) was a UGV that was fully tested several times in the Northern Virginia area. It was learned that SARGE worked, but it had several limitations (e.g., line-of-sight transmission, terrain navigation) and was deemed a good initiative but not ready for use in an AWE.

A disciplined process was established for all experiments. To propagate and standardize the methods for individual experiments, the laboratory developed and published an experimentation procedures manual,[21] which defines terms and sets

[21]Marine Corps Warfighting Laboratory. 2001. *Innovation and Experimentation Processes,* Marine Corps Combat Development Command, Quantico, Va., November 29. Available online at <www.mcwl.usmc.mil/divisions/expplans/i&eprocess.pdf>. Accessed October 7, 2003.

forth experiment procedures, detailing requirements for analysis and conclusions. Each experiment has three basic stages—planning, execution, and analysis.[22] The plans, analysis, and conclusions from individual experiments are comprehensive and are available for viewing through Web-based access.

Environment

The Commandant of the Marine Corps established the environment for experimentation in direct and definitive terms. In his "Commandant's Planning Guidance" of July 1995, General Krulak stated:

> An ongoing program will be established under the CG [Commanding General], MCCDC to serve as the cradle and test bed for the development of enhanced operational concepts, tactics, techniques, procedures and doctrine which will be progressively introduced into the FMF (Fleet Marine Force) in concert with new technologies. . . . This program will serve as the integrating ground for new technologies that we procure or develop with other services. It will provide a focal point for warfighting . . . while allowing me, as the Commandant, to accelerate and direct specific efforts within the process of change. It will be the centerpiece of operational reform in the Marine Corps and will help ensure that emerging technologies for the individual Marine are brought into service expeditiously and effectively. . . . By 1 August 1995, the CG, MCCDC will provide a plan of action for the establishment of this lab. I desire it to be operational by 1 October 1995.[23]

The title of the laboratory, the Commandant's Warfighting Laboratory, left little doubt as to leadership and emphasis. The Commandant provided and articulated his vision for establishing the laboratory and the Marine Corps experimentation program. He emphasized a set of focused goals and objectives and an environment that allowed for failure. He provided for and directed the allocation of resources and set the time line to get the program under way. He also directed command relationships that kept him fully involved. By having the laboratory report through the CG, MCCDC, he ensured easy access to auxiliary resources (modeling and simulation (M&S), Marine Corps University, operation and maintenance support, and so on). This alignment also greatly facilitated the entry of the laboratory's successes into the Marine Corps Combat Development System.

---

[22]As noted in Chapter 2, the Services do not have a standard set of of phases for experiments, although in the aggregate the tasks and functions performed across the cycle of an experiment are similar to those of Marine Corps experiments.

[23]Gen Charles C. Krulak, USMC, Commandant of the Marine Corps. 1995. *The Commandant's Planning Guidance,* Headquarters, U.S. Marine Corps, Washington, D.C., July. Available online at <http://www.usmc.mil/cmc.nsf/CPG?OpenView&ExpandSection=1,2,3,4,5,6,7,8,9,10,11,12,13>. Accessed November 9, 2003.

There was no doubt in the minds of the Marine Corps leadership that the laboratory and Sea Dragon were high on the Commandant's priority list. Within 2 years after the establishment of the MCWL, the need for an experimentation program and the goals and objectives of Sea Dragon were understood and accepted by the majority of Marines.

Organizational Roles and Major Players

As in the past, all Marine Corps experiments are developed and conducted through the MCWL. Most of the experiments are Marine personnel-intensive and, as the MCWL has a very small staff, the majority of the forces used in the experiments came from operational forces, Marines in the training pipeline, and Marines stationed at Quantico, Virginia. To provide continuity of effort during experiments and an operational command element responsible for the actual execution of an experiment, a permanent SPMAGTF Headquarters was organized and staffed as a separate unit of the MCWL. The SPMAGTF Headquarters serves as the experimentation cadre for experimental operations in the Marine Corps. The MCWL coordinates efforts with numerous sources (Federally Funded Research and Development Centers, DARPA, NASA, industry, Service laboratories, academia, allies, and so on) to assist and participate in the Marine Corps experimentation program.

The Marine Requirements Oversight Council (MROC), chaired by the Assistant Commandant of the Marine Corps, and the Marine Advocates, for the support of ground combat, aviation combat, command element, and combat Services, are key participants in the experimentation process. Their participation is linked to the current mechanism for the concept-based requirements system, called the Force Capability Development Phase of the Expeditionary Force Development System (EFDS).[24] All advanced warfighting concepts are briefed for consensus approval to the MROC prior to their final approval. EFDS supports the requirements validation role of the MROC, thereby increasing the ability of the Marine Corps leadership to define, review, and validate the concepts. These concepts constitute the foundation for the Marine Corps experimentation program. The MROC is also briefed annually on the Marine Corps Experimentation Plan (the current version extends to 2008) and on the results of the experiments. During each Executive Offsite (a meeting, called by the Commandant, of top-level Marine Corps officers, held away from the Washington area), the MROC provides Marine Corps leadership with an update on its activities.

[24]The EFDS is the single integrated system of dynamic processes and functions that produces and sustains integrated capabilities which meet the needs of the Marine Corps and the combatant commanders.

Synopsis of Results of Sea Dragon to Date

Both successes and failures were experienced in Sea Dragon. The failures included the remotely controlled urban bulldozer used in Urban Warrior—it was dubbed by the troops "the bulldozer from hell"—the remotely controlled paraglider logistics delivery system, used in Hunter Warrior, that was not ready for use in the field; and the airborne aerostat communications relay system that broke from its tether during high desert winds and threatened Palm Springs, California. However, using the knowledge gained through experimentation, the Marine Corps learned from its mistakes and improved its experimental equipment. Although experimenting with industrial-quality knee and elbow pads for the urban combat troops was unsuccessful, it led to the use of the COTS skateboard equipment that satisfied the requirement.

Another notable success included the development and fielding of the Squad Leader Combat Decision Training Program. This program has been both effective and popular with small-unit leaders and Marines. It is a simulation combat training system that is open-ended, scenario-driven, facilitator-monitored, and computer-projected. The system was developed after visits to and interactions with the New York Stock Exchange, the Federal Aviation Administration, and the New York City Fire Department over 2 consecutive years. Each of these organizations has a unique and thorough training program for its personnel. The MCWL was directed to use the applicable aspects of these programs to develop for small-unit leaders a combat training program, and a prototype was developed in 6 months. Within a year, the basic system, with four scenarios and an associated equipment suite, was fielded to a deploying unit. The fielding of additional systems and scenarios continues today.

The Interim Fast Attack Vehicle (IFAV) Program is another example of success. It introduced a major capability into the operating forces in less than a year. The deficiency addressed was that of a needed augmentation in tactical mobility. Front-line Marine units were deploying jeep fast attack vehicles (FAVs) that were outdated (the last one was produced during the 1970s). Maintenance on these jeep vehicles when based at home was onerous, but when they were deployed it was nearly impossible. The Marine Corps had a major acquisition program to address the problem—the light strike vehicle (LSV) with an initial operating capability planned for 2007. During the UW program, the Marine Corps introduced a series of prototype FAVs to the experimental units. These vehicles represented a vast improvement in performance, reliability, and maintenance over the then-deployed FAVs. As a result of the experimentation, a Fleet Operational Needs Statement was written; it was discussed at the next meeting of the senior Marine Corps leadership, which directed that an immediate solution be provided for rapid fielding. MCCDC, working with the SYSCOM wrote an interim requirement. The SYSCOM responded within 60 days with a COTS solution that included a worldwide maintenance and logistics support program.

Within 6 months, each Marine Expeditionary Force (MEF) took delivery of 20 IFAVs. These are now routinely deployed with all front-line Marine units. The IFAVs performed well in combat during Operation Enduring Freedom in Kandahar, Afghanistan.[25] In the meantime, the light strike vehicle, designed to meet the more comprehensive requirement, continues through the acquisition process.

A recent example of a capability with strong potential for transition to the field is that of the Dragon Eye actually used by the Marines in Operation Iraqi Freedom.[26] This is a UAV that can be transported in an organic backpack; it is intended to support a Marine Corps small-unit leader by providing over-the-hill/over-the-next building surveillance and reconnaissance with real-time day or night video imagery. As a prototype it underwent development and operational evaluation to enable TTP development in 2002. It then was used in the Marine Corps Millennium Dragon '02 experiment to support urban warfare scenarios with a battalion headquarters element. Ten systems have been deployed to Kuwait to support MEF units, with an acquisition scheduled to follow, pending results of this extended user evaluation.

A review of the results of Marine Corps Hunter Warrior, Urban Warrior, and Capable Warrior experimentation campaigns leads to the following conclusions:

- Concepts, doctrine, and TTPs resulting from experimentation have transitioned successfully to forces in the field.
- Experimentation has resulted in changes in minor equipment items in the field.
- Experimentation successes for major equipment items have been very difficult, if not impossible, to transition to fielded capabilities.
- There has been a gradual shift from experimental objectives that address long-term conceptual requirements to those that satisfy short-term operational needs. This shift is due in part to an effort to garner support for the experimentation program within the operating forces that are pressed with immediate deficiencies and that supply the bulk of the experimental force. However, most of the experimentation objectives are tied either to near-term deficiencies affecting operating forces or to long-term conceptual requirements.
- Service experimentation has prepared the Services for joint operations in several areas (command, control, communications, computers, and intelligence (C4I), targeting, and terminal weapon guidance).

---

[25]Committee conversation with then MajGen (Sel) James N. Mattis, USMC, Deputy Commanding General, First Marine Expeditionary Force/Command, General First Marine Expeditionary Brigade, on August 1, 2002.

[26]Jason Ma. 2003. "Experimental Dragon Eye UAV Available to I MEF in Persian Gulf," *Inside the Navy*, Vol. 16, No. 10, March 10, p. 1.

- Marine Corps Tactical Systems Support Activities should be brought in earlier when experimenting with C4I issues and equipment to ensure integration and compatibility with current and future equipment and architectures.
- Because of resource limitations, the scope of FBEs and AWEs should be carefully defined.

Future Plans and Program for Experimentation

The Commandant of the Marine Corps has defined his capstone concept for the Corps in *Expeditionary Maneuver Warfare*,[27] and the *Naval Transformation Roadmap,* with its ties to the 2001 Quadrennial Defense Review, provides a plan to evolve future warfighting capabilities to maximize advantages that are uniquely naval.

For the future, as is the case now, force development and requirements to be determined through experimentation remain the responsibility of the Marine Corps Combat Development Command. The MCCDC has established the Expeditionary Force Development Center to develop concepts, coordinate assessment and experimentation, and integrate the implementation of DOTMLPF across the range of Marine Corps operations.[28] The underlying purpose of establishing the center, however, is to develop a single process—the Expeditionary Force Development System—by which the Marine Corps will be transformed in accord with its capstone concept, Expeditionary Maneuver Warfare.

Access, Sea Basing, and Future Maritime Prepositioning concepts will be core objectives in future experimentation plans. Naval forces offer the Joint Force Commander true expeditionary capabilities—those that not only can move to remote destinations and operate without host nation support or infrastructure, but that also can operate over a sustained period of time without requiring immediate reinforcement. Organic sustainability will be the hallmark of future naval forces.

Naval transformation will support the joint warfighter by delivering new military capabilities that will greatly expand the options available to and under the control of Joint Force Commanders. Inherent in all aspects of this transformation is that naval forces will be committed to and built upon the principles of jointness. Consequently the Marine Corps is aligning its Service experimentation toward a better integration with joint efforts.

---

[27]Gen James L. Jones, Commandant of the Marine Corps. 2001. *Expeditionary Maneuver Warfare: Marine Corps Capstone Concept,* Warfighting Development Integration Division, Marine Corps Combat Development Command, Quantico, Va., November 10.

[28]Col Frank DiFalco, USMC, Joint Concept Development and Experimentation Operations Center, Marine Corps Combat Development Command, "Marine Corps Role in JCDE," presentation to the committee on August 15, 2002.

The Chairman of the Joint Chiefs of Staff Instruction 3010.02A dated April 15, 2001,[29] implemented the Joint Vision Implementation Master Plan (JVIMP), consisting of three closely related processes: (1) Joint Concept Development, (2) Joint Experimentation and Assessment, and (3) Joint Integration and Implementation. New joint operational concepts will be assessed through experimentation and assessment activities and will be developed with formal Service Headquarters, Combatant Commanders, Joint Staff, Joint Warfighting Capabilities Assessment (JWCA) teams, and selected Office of the Secretary of Defense (OSD) agencies for coordination.

Joint Experimentation and Assessment activities evaluate alternatives in order to achieve desired operational capabilities and articulate results in terms of recommended changes to joint DOTMLPF. The Joint Integration and Implementation component initiates the process for integration and the implementation of recommended changes. This process is consistent with USJFCOM's role as the DOD lead for transformation, while the Joint Staff provides the guidance for future force design.

The Assistant Commandant addressed these process changes during MROC meetings. He directed that the Marine Corps improve its responsiveness to the JVIMP and that the Marine Corps more fully engage with USJFCOM and the Joint Staff through the JWCA process. He further stated that the MROC must oversee Marine Corps participation, and he identified MCCDC as the USMC lead in this effort. Additional instruction from the MROC meetings directed a better integration of Marine Corps combat development at all key JVIMP junctions. Also, the Marine Corps was directed to validate its capabilities in the joint context and to ensure that Title X responsibilities were part of the joint equation.

To accomplish these tasks, the CG, MCCDC, was directed to establish a Joint Concept Development and Experimentation Office (JCDE Office). He, in turn, set up his lead JCDE office at Quantico, Virginia, with branch offices at the Pentagon and at Suffolk, Virginia, colocated with USJFCOM. The published mission of the Marine Corps JCDE Office is "to integrate the Marine Corps force development process into the Joint force development process in order to provide Marine Corps capabilities for the future Joint Force Commander."

Many of these recent actions are intended to aid and accelerate the transformation of U.S. forces. The Marine Corps's next major experiment, Olympic Dragon 2004 (OD04), will be conducted within the context of the U.S. Joint Forces Command's Rapid Decisive Operations integrating concept. OD04 is examining the art, not just the science, of command and control. The focus of OD04 experimentation is on the people and information associated with com-

[29]Joint Chiefs of Staff. 2001. *Joint Vision Implementation Master Plan (JVIMP),* Chairman of the Joint Chiefs of Staff Instruction (CJCSI) 3010.02A, The Pentagon, Washington, D.C., April 15.

mand and control, not only on the hardware and associated systems. This is an essential step, because the Marine Corps cannot execute Ship-to-Objective Maneuver as envisioned with currently planned command and control information technology capabilities.

COMBINED EXPERIMENTATION OF THE NAVY AND MARINE CORPS

In the past, the Navy and Marine Corps have aligned their concepts selectively. Several advanced warfighting concepts—such as the Concept for Future Naval Mine Countermeasures in Littoral Power Projection and the 1998 concept Sea-Based Logistics—have been developed, signed, and published jointly by the Commander, Navy Warfare Doctrine Command, and the Commanding General, Marine Corps Combat Development Command. For the future, to ensure that both Services continue along the prescribed path of the *Naval Transformation Roadmap* together, permanent reciprocal billets have been established and filled at both the NWDC and the MCCDC. Two key positions have been given responsibility for two assignments: that of the CG, MCWL, as the Vice Chief of Naval Research; and that of the Chief of Naval Research as the Deputy Commandant for Programs and Resources on the Marine Corps staff. Such reciprocity is intended to solidify the collaboration of both Services in their work toward naval transformation.

There are numerous examples of combined efforts of the Navy and the Marine Corps in experimentation. The numbered fleet commanders normally conduct FBEs, usually in conjunction with training exercises (e.g., Kernel Blitz), AWEs, or carrier strike group (CSG) or ship or unit certification events. The Navy and Marine Corps each have training, certification, or experimental objectives in combined FBEs and exercises. During FBEs, the majority of the experimental objectives are Navy objectives. During AWEs the majority of the experimental objectives are Marine Corps objectives. Due to the large number of assets required and the operational and personnel tempo of units and people,[30] both Services frequently align various experimentation objectives with advanced warfighting concepts and near-term operational requirements.

Many of the FBEs have involved the efforts of both Services. Several Marine Corps experimental objectives were included and worked on as part of Navy FBEs and, in turn, several Navy objectives were included in Marine Corps AWEs, but in neither case were these objectives highlighted in the list of objectives appearing in the descriptions of the larger-scale experiments. These were primarily LOEs and LTAs. One such example addressed the long-term problem associ-

---

[30]Operational tempo refers to naval units, and personnel tempo refers to people. "Tempo" means the duration and the frequency of overseas deployment.

ated with the storage of "Mogas" (commercial gasoline) aboard amphibious ships for use in Marine tactical vehicles. This issue was temporarily resolved when the Marines experimented with various Interim Fast Attack Vehicles through an LTA discussed above. Another example involved addressing battle casualty treatment through telemedicine; the successful experimental objective included a separate LOE of FBE-E and an LOE of AWE Urban Warrior, as well as the award-winning Multipurpose Health Service Facility or "doc in a box" (winner of a 1998 *Popular Science* magazine annual competition).[31] The experimental use of the SLICE ship to enhance rapid combat supply capability for troops ashore is another example of an LTA of FBE-E that combined Service objectives.

During the early stages of FBE-E and Urban Warrior, the Navy and the Marine Corps focused primarily on Service objectives. During the final stages of the same experiments, the Marines moved aboard the ships involved in the FBE, and the Services combined efforts to focus on naval objectives—to experiment with various shipboard command and control systems that would maintain contact and provide information to ground forces as they moved through urban terrain. The systems were to provide a common operating picture for all users, provide automatic intelligence updates, provide information on the supply and resupply status for all units, and guide precision fires.

Planning and execution for this last stage were both energetic and enthusiastic. To further explore the expanding use of naval forces, a humanitarian and disaster relief experiment was conducted. In order to provide realistic scenarios and the kinds of personal interactions associated with these types of missions, a wide range of new participants (actual city civilian officials and agencies) as well as surrogate facilities (Naval Postgraduate School) were incorporated in the experiments. The learning curve was steep for all participants, but the procedures and protocols for civil and military interaction in humanitarian missions were greatly advanced.

Doctrinal changes in command relationships were also explored. Following the Kernel Blitz exercise, the traditional command relationships of Commander, Amphibious Task Force, and Commander, Landing Force were replaced with the supporting and supported relationships normally associated with joint operations. This successful experiment greatly assisted the recent changes in naval doctrine regarding command relationships. Such naval experimentation partnerships have resulted in many innovations, including emerging telemedicine, improved ship-to-ship communications, a single fuel for deployed USMC vehicles, several command relationship options, and the first-time deployment of the Dragon Drone with the MEUs and ARGs in the late 1990s.

---

[31]Sgt Jason Bortz, USMC. 1999. "Doc-In-A-Box: New Tent Means Better Medicine for Combat Marines," *Marines* Magazine, Vol. 28, No. 1, January, p. 13.

Many of these capabilities and changes have been adopted by operational units and have resulted in changes to DOTMLPF, but to date, enhancements to equipment have been few and on a small scale. No major programs of record have been changed or initiated.

Both the Marine Corps and the Navy participated in the Extending Littoral Battlespace ACTD. This multiyear effort was designed to help develop and distribute a common operational picture to units afloat and ashore and to reduce the number of nodes required for communications and data transfer. Although surrogate equipment was used and the system was unclassified, the system did pass video, data, and voice transmissions. Both the Navy and the Marine Corps users asked to keep the equipment and to experiment during their upcoming deployment. However, no new equipment or capabilities have been developed or purchased for operational units owing to the lack of secure transmission, the immature nature of the technology, the amount of developmental funding required, and the lack of transition funding and authority.

EXPERIMENTATION BY OTHER SERVICES

The committee held discussions with representatives of the Army and the Air Force to acquire an understanding of how these Services approach and use experimentation for the development of their future military capabilities. The committee's goal was not to assess these other Services' experimentation activities but to extract ideas and lessons learned that could benefit the Navy and Marine Corps in their future experimentation efforts.

U.S. Army

Overview

U.S. Army experimentation was reenergized in 1956 with the formation of the Combat Development Experimentation Command (CDEC). The mission of CDEC was to conduct experiments to discover how to make the Army more effective. The experiments were primarily field exercises, with umpires overseeing the operations and trying out new ideas to see how they worked. Such ideas included evaluating operational concepts for unmanned aerial reconnaissance vehicles as early as the 1970s. CDEC evolved into a technology-focused command in 1970, with the development of an instrumentation system that controlled experiments and collected data automatically. It continued this mission of combat development until becoming an operational test command in about 1980. From 1980 until its dissolution, it remained centered on operational testing but continued to do some combat development-type experimentation.

Today combat development experimentation is done under the aegis of the Training and Doctrine Command (TRADOC). In addition to its training mission,

TRADOC oversees concept development along with new doctrine, tactics, techniques, and procedures. However, experimentation is executed mainly by individual battle laboratories that are associated with individual branches of the Army. For example, artillery experimentation is done at Fort Sill, Oklahoma; infantry experimentation, at Fort Benning, Georgia; and air defense experimentation, at Fort Bliss, Texas.

Various members of the Army research, development, and acquisition community sometimes participate in experimentation. While their primary mission is centered on moving individual acquisition programs through the required processes to achieve full-scale production, program managers are interested in experimentation to the extent that specific systems require or enable new operational concepts or TTPs. These individuals have also provided specific capabilities for experimentation activities, such as for LOEs and AWEs, as prominently illustrated by the Army's experimentation campaign for digitization (discussed below). The Army's Research, Development, and Engineering Centers sponsor prototypes or provide surrogate capabilities for experimentation activities.

A Past Example of Army Experimentation

The use of experimentation to achieve digitization was initiated in the early 1990s when then Chief of Staff of the Army General Gordon Sullivan realized the need to shift force capabilities to exploit information technology. His objectives were increased survivability, lethality, and operational tempo resulting from significant improvements in situational awareness. The long-term goal was for every person in the Army to know where he or she was and where the enemy was—and to achieve that goal with some form of fielded capability by 2010. This goal required a substantial change not only in DOTMLPF but also in the processes by which acquisition was accomplished in the Army. In short, achievement required a spiral process, which at the time was viewed as the antithesis of the Army's standard linear approach of capturing detailed requirements "in stone," sending them off to various contractors for development, and getting systems delivered approximately 14 years later.

General Sullivan assigned General William Hartzog to TRADOC in 1994 to achieve these objectives. He also made clear to the entire Army leadership the priority that he placed on the digitization goal. A plan to communicate its importance throughout all the Army was developed and carried out. The Army developed an Experimentation Campaign Plan with a view to developing a division-level prototype, and Army senior leadership maintained oversight of progress. To coordinate and integrate efforts of the various participating organizations, General Sullivan also established the Army Digitization Office. The office reported directly to him.

Experimentation served several objectives with respect to the digitization goals, but one of these was to determine what capabilities should be fielded and to accelerate that process. In response to the ambitious fielding dates, the Army applied spiral processes of all four types (see Box 2.1 in Chapter 2). It spiraled concepts to understand the implications of digitization. It spiraled through a series of experimentation campaigns built around what was learned about digitization at company, battalion, brigade, and division levels. For each of these unit experiences, the Army spiraled through a spectrum of activities, including games, modeling and simulation, LOEs, field exercises, and AWEs, to gain an understanding of required capabilities. It also applied spiral development to field an ambitious brigade-level architecture[32] linked to the Central Technical Support Facility at Fort Hood, Texas; used a special oversight management structure; and designated the Fourth Infantry Division as an experimental cadre.

In March 1997, the Army conducted a brigade-level AWE, called Task Force XXI, at the National Training Center to evaluate the effects of digitization on lethality and operational tempo. It used a force-on-force encounter with 15,000 entities on the "battlefield" and maintained situational awareness on them. Included in the experimental capabilities were 71 initiatives requiring 66 new books of doctrine, used by the forces in the experiment. Following the event and its evaluation, eight new pieces of equipment were validated and designated to be acquired immediately; the Army subsequently received $100 million in funding for a rapid acquisition program for selected small items outside the normal annual authorization and appropriation process. Another important outcome was that the Army Chief of Staff decided to digitize the III Corps.

A review of these digitization efforts indicates that the five key factors for experimentation to achieve successful change were present (see subsection entitled "Experimentation That Changed Operational Capabilities," earlier in this chapter). The Army had and/or provided the following:

1. A problem to be solved—a compelling need;
2. Relevant technology;
3. Leadership buy-in (starting with the Chief of Staff of the Army);
4. Organizational structure (the assignment of General Hartzog, the designation of an experimental cadre, and the use of special management structures for oversight and integration); and
5. Funding.

Other lessons from these ambitious experimentation campaigns can be derived.[33] They include the need for the following:

---

[32]See BG Steven Boutelle, USA, and Alfred Grasso, 1998, "A Case Study: The Central Technical Support Facility," *Army RD&A* (now *Army Acquisition, Logistics & Technology (AL&T) Magazine*), March-April, pp. 30-33.

[33]Based on committee discussions with GEN William Hartzog, USA (retired), President and COO, Burdeshaw Associates, on May 3, 2002.

• Senior involvement (four-star rank is necessary; two-star rank is insufficient): The most senior leadership (Chief of Staff) must believe in experimentation and drive it;

• A risk culture (mistakes will be made);

• Sustained leadership to manage experimentation campaigns (a minimum of 4 years);

• Industry incentives (to compete for products already in the program of record);

• Testing embedded within the experimentation process (technologically successful architectures may not be operationally joint);[34]

• Sufficient time for the assimilation of results (the time required can span as much as 5 years);

• Mechanisms for the capture of knowledge (knowledge needs to be codified);

• Sufficient funding (substantial funding should be planned for);

• Communications plan (to relate experiments to the common vision and to broadcast results to stakeholders); and

• Early training of operators on new capabilities to evolve the necessary business rules for their use.

The Army also gave great credit for its success to the Central Technical Support Facility, an integration and testing facility that had as its mission "to act as an enabler for rapid integration of dissimilar software and hardware systems through real time interaction with soldiers, contractors, testers, program managers, and the requirements community."[35]

A postscript is warranted regarding the results of digitization experimentation. Acquisition programs were displaced and new programs were added. One critical factor, that of funding to acquire advanced capabilities, was constrained. Only three major new acquisitions were actually realized—the Force XXI Battle Command Brigade and Below System (FBCB2); the Army Battle Command System, which included a whole family of new command and control systems; and the Army's tactical internet, which included the Warfighters Information Network–Tactical, line-of-sight tactical radios, and beyond-line-of-sight communications.

An entire division (the Fourth Infantry Division) was digitized, as was the current Stryker Brigade at Fort Lewis, Washington. Tactical communications—both data and voice—were completely changed to accommodate Internet Proto-

---

[34]The Army test community was made part of digitization experimentation. Today the U.S. Army Test and Evaluation Command exercises a role in experimentation different from that of its traditional operational test and evaluation responsibility for acquisition.

[35]See BG Steven Boutelle, USA, and Alfred Grasso, 1998, "A Case Study: The Central Technical Support Facility," *Army RD&A* (now *Army Acquisition, Logistics & Technology (AL&T) Magazine*), March-April, p. 30.

col networking. C2 and ISR programs of record were changed. The impact of the new programs was widespread, affecting many weapons platforms, most notably the M1A2 System Enhancement Program (modernized Abrams tank), M2A3 (Bradley fighting vehicle), aviation, and multiple-launch rocket system. These impacts were technical (such as requiring that systems use components of FBCB2, scheduling (delaying some key milestone events such as the M1A2 initial operator test and evaluation to ensure synchronization with FBCB2), and funding (requiring additional funds for these programs to make them compliant with digitization requirements and strategy). The linkage between the success of FBCB2 and its impact on the interfacing systems for the Future Combat System was a difficult challenge that has had a broad impact on Army acquisition.

Current Activities and Future Plans

Today the Army has shifted its strategy from its previous approach of digitization of heavy forces. In 1999 the Army acknowledged an inability to deploy the force rapidly to compel decisive outcomes. The current challenge for the Army is to provide a lethal force that can be decisive more quickly than its current light brigades and/or light divisions and that can be deployed more rapidly than its current heavy armored force.

The Army's target for transformation is its future Objective Force (OF), which over the next several decades will provide new capabilities for each Combat Service and Combat Service support area, especially for armor and infantry. Already-fielded capabilities, including those from digitization, will be maintained through the Legacy Force. The OF roadmap, which began in 2000, involves three distinct paths: the Legacy Force, an Interim Force, and the Objective Force.

The operational concept for the OF is to employ an unprecedented level of C4ISR, enabling leaders at all levels to exercise initiative within the commander's intent to dictate the time and place of engagements on future battlefields. The OF will also involve new approaches for echelons. OF units of action will include combat units that will be comparable to today's brigades, whereas OF units of employment will interface with joint commanders and theater systems and be comparable to today's divisions.

The Interim and Objective Forces specifically address newly acknowledged operational challenges—most importantly, the rapid deployment of a brigade in days. For the Interim Force, six brigades will be equipped with new "digitized" medium-weight vehicles that can be deployed through airlift. The Interim Force is structured around the Interim Brigade Combat Team and is equipped with the Stryker wheeled vehicle, which will have 10 variants, employ the latest off-the-shelf vehicle technologies, and owe all of its C4ISR capabilities to digitization efforts. However, these vehicles do not have the lethality or survivability of today's Abrams and Bradley vehicles.

The OF will receive substantially new capabilities resulting from aggressive science and technology efforts and fast spiral developments. The Future Combat System (FCS) includes 19 new vehicle variants and C4ISR systems for maneuver units of the OF. C4ISR integrates the FCS into a system-of-systems, affording greater lethality and survivability than the current Abrams and Bradley vehicles while supporting the rapid deployment capability of the Stryker. The Army has set a goal of achieving an initial operating capability for the FCS by 2010, when other OF systems, such as Force Warrior, Comanche, Warfighters Information Network–Tactical, and the High Mobility Artillery Rocket System, will be fielded. An architecture to subsume all of these systems is still under development.

Future Army Experimentation

Experimentation is important to the success of both the Stryker and the FCS efforts, but in different ways. For the Stryker, key issues revolve around developing the TTPs. In the case of FCS, there are significant changes in the unit-of-action concept of operations that warrant experimentation. Because there are limited surrogates for these new capabilities, much of the experimentation will of necessity rely on virtual or constructive modeling and simulation.

For the Stryker, the Army is using the traditional linear acquisition approach. However, the first versions of the Stryker that were produced (before full-scale production approval, which is yet to occur) did engage in the Millennium Challenge '02 experiment managed by USJFCOM. This event provided the opportunity to evaluate the effectiveness and suitability of the Stryker, while developing doctrine and TTP before it enters its 18-month period of operational testing and evaluation to determine its suitability for full-scale production.

The production goal for equipping six Stryker Brigade Combat Teams (SBCTs) is about 2,100 vehicles. The goal for certification to "go to war" after the completion of operational testing and evaluation is 2003. It is not known exactly how and when the Army plans to use the experimental results to decide upon and then make adjustments and/or improvements to the Stryker for this accelerated acquisition program. At the request of OSD, the Army is currently evaluating potential improvements to equipment for the fifth and sixth SBCTs, which may be fielded through FY 2004 to FY 2009.

For the Army's FCS, digitization experimentation has provided lessons learned, but the C4ISR is being designed with a "clean sheet of paper." The strategy for FCS is to allow the design to spiral until 2005, but the schedule for its initial operating capability stands firm at 2010. The Army has adopted an innovative approach to acquisition called other transaction authority, to be used in conjunction with spiral development. Other transaction authority enables a cooperative relationship with industrial partners. The Army, teamed with DARPA, has selected a lead system integrator—a team of contractors—in order to seek out and integrate the most promising and cost-effective technologies. From the perspective of requirements, design, and experimentation, the important elements are that the user requirements will evolve in the initial development process as the contractor team, DARPA, and the Army learn from the spiral development process, including experimentation and risk management with continuous feedback. In this sense the FCS development is different from that for the Stryker family of vehicles.

The Army will use the Stryker capabilities to build the experiential base required for the initial OF units of action. A program of experimentation has been established at Fort Knox, Kentucky, to address unit-of-action experiments. This facility, which builds on the efforts of the Mounted Maneuver Battle Laboratory, is known as the Unit of Action Battle Laboratory. It is conducting a number of force-on-force events to explore how the unit fights and how it will be optimally organized. It has also conducted a C4ISR experiment.

The Army anticipates that FCS will become an acquisition program in May 2003. To provide evidence for FCS acquisition decisions, very small, focused events have been conducted to prove that operational and system concepts are sound. However, none of these events has been on the same scale as that of the AWEs conducted for digitization. They have actually been called demonstrations to highlight differences in the types of events. Demonstrations at military facilities have involved roughly 20 surrogate FCS manned ground vehicles (e.g., sport utility vehicles and Legacy Force vehicles) and one or more surrogate FCS unmanned aerial vehicles (e.g., Blackhawk helicopters), or virtual simulations. It appears that formal experimentation is not ongoing, at least while FCS is in a pre-acquisition state. Experiments for FCS or the OF on the magnitude of an AWE are not currently included in campaign plans or schedules leading to initial operating capability.

For FCS, the Army's use of experimentation differs from what it did for digitization. The past cost of experimentation measured in dollars and human resources may be a factor, particularly that associated with large-scale events. However, the difference is not merely one of scale. There is yet no Army experimentation campaign plan for this, although there are efforts within TRADOC to develop what might be called an objective force experimentation campaign plan and to ensure that these Army activities are synchronized with the USJFCOM plans for experimentation. Also there is no special senior management structure in place (beyond TRADOC) to review OF experimentation.

Both Stryker and FCS are being managed as acquisition programs, using experimentation as a means to accelerate the acquisitions or determine relevant parameters; they are not part of an Army experimentation campaign to investigate the most desirable directions that the Army might pursue in developing or improving future forces. Said differently, the Army has determined what it needs and wants in its future force, and it will use experimentation to get there. However, it has not set up an experimentation campaign to provide and evaluate multiple alternative means of improving its future force. The latter option would have taken more calendar time, and conceivably the decision has been made explicitly or implicitly not to accept such a delay. The result is the acceptance of an increased risk and the desire to use experimentation and testing in the spiral development and block procurement processes to manage such risks.

U.S. Air Force

Overview

The U.S. Air Force (USAF) has a long tradition and culture of experimentation, ranging from the 1920s, when General Billy Mitchell attacked an ex-German battleship in the Chesapeake Bay to investigate air search and air power, to today's experiments that explore stealth-delivered, precision-guided weapons of the future.

The Air Force describes the purpose of its experimentation programs as follows: ". . . to explore new operational concepts and technologies that will provide the capabilities to achieve [the Air Force] vision. . . ."[36] There is published policy and guidance on how experimentation is conducted within the Air Force.[37] That guidance expressly intends the results of experimentation to inform investment and divestment decisions, identify transition candidates for fielding, and validate changes to DOTMLPF. In short, the objective of experimentation is to quickly and efficiently improve the concepts, processes, and systems associated with air and space warfare capabilities at the strategic, operational, and tactical levels.

Experimentation, as managed by the Air Force, includes a full spectrum of activities, from tabletop strategic war games to detailed, tactical, human-in-the-loop events that are highly instrumented. Among other activities, it includes ACTDs, large-scale Service experiments such as the Joint Expeditionary Force Experiments (JEFXs),[38] small-scale experiments such as the Advanced Process and Technology Experiments,[39] and experimentation events combined with field and command post exercises. It also includes participation in joint experiments.

The USAF candidates for experimentation come from numerous sources. These include organizations such as its Major Commands (MAJCOMs) and battle

---

[36]Lt Gen Robert H. Fogelsong, USAF, Deputy Chief of Staff, Air and Space Operations. 2000. *Air Force Experimentation Campaign Plan FY00-05,* Department of the Air Force, Langley Air Force Base, Va., p. 3.

[37]AFI 10-2304 was still in draft form as of this writing.

[38]Equivalent to an FBE or an AWE.

[39]Equivalent to an LOE or an LTA.

laboratories, various events such as its Blue/Red/Green Flag battle management and flying training warfighting exercises and JEFXs, and from various individuals such as the senior USAF warfighters, such as Global Strike Task Force (GSTF), Multi-Mission Command and Control Constellation/Aircraft, and others. Concepts originate from any of the six Air Force staff task forces, from those involved in working the main elements of the Air force vision, and from various related sources.

Experimentation campaigns are documented in an Air Force experimentation campaign plan that is tied directly to the Joint Vision, to the Service Vision, and to the *Air Force Transformation Flight Plan*.[40] There is also a management structure and a well-established process within the USAF to oversee, plan, and execute USAF experimentation and to make the resulting changes in doctrine and procedures.[41] This process is well supported by the senior Air Force leadership. Results of experiments are reviewed first by a Council of Colonels, and then by a General Officer Steering Group for review and appropriate action. Recommendations may be briefed to the Air Force corporate structure, including the Chief of Staff. These briefings are usually provided after every major experiment.

Organizational Roles and Major Participants

The responsibility for executing USAF experimentation generally rests with the MAJCOMs and their associated battle laboratories, but there are many important participants, such as the Air Force Doctrine Center[42] and the Air Force Research Laboratory, to mention just two. The organization and conduct of USAF experimentation is somewhat decentralized, in order to accommodate not only distributed responsibilities but also initiatives that arise from many sources. The appropriate MAJCOM oversees its assigned battle laboratories, while another part of the MAJCOM looks after Blue/Red/Green Flag battle management and flying training warfighting exercises. Yet another organization handles major command and control experiments, such as JEFX, under what is called the Air Force Experimentation Office[43] (AFEO). And finally, ACTDs and other experiments directed by senior USAF officers are assigned to appropriate MAJCOMs

[40]Gen John P. Jumper, USAF, Chief of Staff, and James G. Roche, Secretary of the Air Force. 2003. *Air Force Transformation Flight Plan FY03-07,* Headquarters, U.S. Air Force, Washington, D.C.

[41]Lt Gen Robert H. Fogelsong, USAF, Deputy Chief of Staff, Air and Space Operations. 2000. *Air Force Experimentation Campaign Plan FY00-05,* Department of the Air Force, Langley Air Force Base, Va.

[42]The Air Force Doctrine Command has the responsibility of maturing experimentation results into Air Force doctrine after an extensive approval process.

[43]This is something of a misnomer as it does not handle all USAF experimentation but rather major command and control experiments only.

for execution. The development and execution of USAF experiments are organized under the operating side of the USAF. The USAF acquisition community, represented by the Air Force Materiel Command and its operating centers, is closely associated with all USAF experimentation activity so as to facilitate transitioning into acquisition those capabilities that may come from the experimentation process.

The function within the Air Force that brings all USAF experimentation output into focus and coordinates its transition into acquisition as appropriate is the USAF requirements/capability acquisition process. As various experiments from their sponsoring organizations are designed and executed, they are reviewed by the appropriate MAJCOM and Headquarters USAF Requirements Oversight Council. The Service equivalent of the JROC, it is called the Air Force Requirements Council (AFROC); it is analogous to the MROC, discussed earlier in the chapter. The outputs of the USAF experimentation process that are found worthy of becoming USAF capabilities are entered by the sponsoring MAJCOM or senior officer into the USAF requirements process to compete for the funding necessary for the acquisition process to proceed.

Many organizations foster and participate in individual experimentation activities. For instance, a JEFX has the Air Combat Command (ACC) as the executive agent, with the AFEO as the planning lead, Electronic Systems Command as a technical lead, and another ACC component as the operational lead. In addition, many other organizations may participate, depending on the nature and subject of the experiment. The Air Force experimentation campaign plan is a means of coordinating their activities. Examples of other participants are the exercise and training facilities, such as those at Hurlburt Field, Florida, and Nellis Air Force Base, Nevada; the Air Force Test Agency; and program offices, which sponsor initiatives such as UAVS and unmanned combat air vehicles (UCAVs) as subjects for experiments. Collectively, these participants may bring a full family of advanced engineering, engagement, mission, and campaign models and simulations as well as provide facilities, unique infrastructure, and specialized tools that make up the essential support for experimentation activities.

Specific organizational participants in USAF experimentation are as follows:

- *Air Force Experimentation Office.* The AFEO manages experimentation within the USAF having to do with command and control and with intelligence, surveillance, and reconnaissance. Its most notable experiments are those associated with the Joint Expeditionary Experiments in 1998, 1999, 2000, and 2002 (discussed below). The AFEO, which is part of the USAF C2 and ISR Center, was established in 1998 under the Air Combat Command. In recognition of its importance to the USAF as a whole, the C2 and ISR Center, along with the AFEO, was recently realigned under the Chief of Staff of the USAF.
- *USAF battle laboratories.* Today the USAF has seven Air Force battle laboratories, each having fewer than 25 personnel, commanded by an O-6 (rank

of colonel), and sharing equally in the funding available for experimentation.[44] The command and control battle laboratory reports to the Air Force C2 ISR Center, the Space Battle Laboratory is under Air Force Space, and all the others report to the ACC.

The battle laboratories receive and/or generate ideas that are screened and selected to be initiatives for experimentation. With some exceptions, it takes from 3 to 6 months for an idea to be approved as an initiative; each initiative has an 18-month execution phase. Consequently, the lifetime of an initiative is on the order of 2 years and, if successful, the initiative should transition into an operational capability. Major initiatives are labeled "Mitchell" and more modest ones, "Kenney." Mitchell initiatives have turned out to be beyond reach, so that all 90 initiatives[45] were of Kenney status. As of the summer of 2000, of 90 initiatives, 40 were still being worked on, and 50 had been completed (all 50 are listed in Table 3.2).

The seven USAF battle laboratories are these:

- The Air Expeditionary Battle Laboratory at Mt. Home Air Force Base, Idaho;
- The Command and Control Battle Laboratory at Hurlburt Field, Florida;
- The Unmanned Air Vehicle Battle Laboratory at Eglin Air Force Base, Florida;
- The Space Battle Laboratory at Schriever Air Force Base, Colorado;
- The Force Protection Battle Laboratory at Lackland Air Force Base, Texas;
- The Information Warfare Battle Laboratory at Kelly Air Force Base, Texas; and
- The Air Mobility Battle Laboratory at Scott Air Force Base, Missouri.

Certain facilities for exercises, training, and testing offer specialized capabilities to support experimentation. For example:

- *The Hurlburt Field, Florida Blue Flag Training Facility, Florida.* The centerpiece of the laboratory is its modeling and simulation capability—it can provide a dynamic and realistic backdrop for both training and experimentation for USAF Senior Leader Operational Battle Command and Control.
- *The Nellis/Edwards Air Force Base/Fort Erwin Western Test and Training Complex, California.* This range complex is used to train aircrews at the tactical level while they interface with operational command and control in red/green flag exercises and large-scale joint exercises and experiments, the most recent being Millennium Challenge '02.

---

[44]Typically about $5 million per battle laboratory.

[45]As of the summer of 2000.

TABLE 3.2 Battle Laboratory Initiatives Completed as of Summer 2002

Battle Laboratory	Initiative	Status
C2B	Hill ATO Defragger	F (Kosovo, USAFE, 7 Wings at ACC)
AEFB	Integrated Planning and Execution Capability	F (Kosovo), R (Crisis Action System)
C2B	Collaborative Tools	F (Kosovo), R (AC2ISRC)
FPB	Ground Based Radar	F (3 sites, South America)
SB	Site Protection (Operation Geese)	F (55th SWS)
FPB	Space Environment	F (820th SFG)
IWB	Network Display Sensor Guard	F (AFCERT)
C2B	Network Attack Visualization	F (AFSOC, F117, 8th AF, 1st MEF)
FPB	Reduced Hardware Footprint	F (CENTAF, CENTCOM)
IWB	Food and Water Antiterrorism	F (Cheyenne Mountain, FBI)
IWB	Enhanced SA Tool	F (classified customer)
IWB	Diagnostic Emulator	F (classified customer)
C2B	Voice Optimal Interrogation	F (GCCS, TBMCS 1.0.2)
IWB	Enhanced Linked Virtual Information System	F (GPS JPO NAVWAR Tool)
C2B	Miniaturized GPS Jammer	F (Master Air Operations Planner)
SB	ATO Visualization and Assessment	F (Operation Northern Watch)
AEFB	Commercial Applications for Combat Effectiveness	F (Residual Cap: 2 KC-135Rs M)
IWB	EOC Enroute	F (SWC Red Team)
IWB	Software Agent System for OPSEC Information Warfare (SCI) Reachback	F (USAF, USA)
FPB	Vehicle Entry Explosives Search Strategy	T (guide published)
C2B	Tactical Sensor Integration	T (ASOC and BCC)
SB	Hyper-spectral Imagery Collection Upon Pike's Peak	T (training changed)
IWB	Signal Analysis Mapping	P (ACC current purchase)
AEFB	Combined AGE	P (IOC Nov. '01)
AEFB	Compact Air Transportable Hospital	P (AF/SG-buying~100 per year)
SB	Space Surveillance Network Optical Augmentation	A ('02 POM candidate)
UAVB	JSTARS Battlespace Imaging	A ('05 implementation)
FPB	Remote Visual Assessment	A ('02 POM vandidate)
FPB	Pathogen Indent Device	A ('02 POM)
AEFB	Next Generation Munitions Trailer	A (ACC/DRW writing ORD)
IWB	Pulse Doppler Identification	A (ECM Pods)
AEFB	Deployment Personnel Accountability Readiness Tool	A (Joint ORD)
C2B	Speech Recognition	A (AC2ISRC)
IWB	Network Early Warning	A
IWB	Cyber Warrior	A
IWB	Re-configurable EW Avionic Parts	A
SB	Space Object Indent in Living Color	A

continues

TABLE 3.2 Continued

Battle Laboratory	Initiative	Status
UAVB	FAA Airspace UAV TCS	A
IWB	Panther Den	A/Masked
C2B	TBMCS and ABCS Data Sync	Further research
SB	Satellite Track Using Ambient RF	Further research
UAVB	SEAD Enhancement	Further research
UAVB	Communication Relay	Further research
UAVB	Spotter UAV	Further research
AEFB	Common Bore Sight	Not recommended
AEFB	Harvest Phoenix	Cancelled (redundant)
FPB	Virtual Tower	Cancelled
SB	Space Doctrine	Cancelled (not meeting objectives)
C2B	JFACC Project Phase 2	Technology not mature
FPB	Hazard Assessment and Mission Enhancement of Resources	Technology not mature

NOTE: F = fielded; T = changed; A = awaiting acquisition process; P = in Program Objectives Memorandum (POM). SOURCE: Lt Gen Robert H. Fogelsong, USAF, Deputy Chief of Staff, Air and Space Operations. 2000. *Air Force Experimentation Campaign Plan FY00-05,* Department of the Air Force, Langley Air Force Base, Va., p. 20.

- *The Eglin Air Force Base Land and Water Range Complex, Florida.* This range complex is used to test, train, and experiment with the full spectrum of live air-to-air and air-to-ground precision-guided weapons.
- *The Combined Air Operations Center-Experimental (CAOC-X), Virginia.* This center was established about 3 years ago to support experimentation with processes, procedures, and systems associated with the USAF Air and Space Operations Center.[46] It was intended to facilitate the acquisition of fielded capabilities through a rapid spiral process, resulting in "leave behinds" for operations.[47] However, owing to resource constraints and the operational urgency of establishing an updated CAOC at Prince Sultan Air Base in Saudi Arabia, the objectives for CAOC-X appeared in flux at the time of this study. Nonetheless, CAOC experimentation has proceeded under the supervision of the AFEO using the facilities and infrastructure at Hurlburt Field and Nellis Air Force Base.

Many other facilities provide extensive modeling and simulation capabilities, and/or offer testbeds. For example:

[46]Analogous to the U.S. Army's Central Technical Support Facility at Fort Hood, Texas.

[47]Naval Studies Board, National Research Council. 2000. *Network-Centric Naval Forces: A Transition Strategy for Enhancing Operational Capabilities,* National Academy Press, Washington, D.C., Section 2.5.4.2.

• *The Paul Revere Boeing 707/767* flying experimental testbeds for the Multi-Mission Command and Control Aircraft (MC2A).

• The THUNDER, BRAWLER, SUPPRESSOR, Joint Simulation System (JSIMS), Joint Warfare System (JWARS) family of campaign, engagement, and mission simulations, with their numerous supporting engineering models, are key capabilities for experimentation.

• *The Air Force Operational Test and Evaluation Center (AFOTEC)* is responsible for the operational testing of new systems being developed for the Air Force and for multi-Service uses. It is a directorate, reporting directly to the Chief of Staff of the Air Force. It has become involved in development evaluation and is now involved at the beginning of experimentation. For instance, for an ACTD, AFOTEC will do an operational assessment, but it is not the usual test before the acquisition of a production capability. It is instead an assessment that goes to the developer for review. If the experimental capability is successful, AFOTEC evaluates the initiative later, after it has evolved through further experimentation. At the appropriate time, AFOTEC conducts the usual test and evaluations that support the formal acquisition process. AFOTEC's earlier development assessment is "nonthreatening," serving to point out issues that need resolution through experimentation and identifying key problems such as critical safety issues, and it familiarizes the test community with how operators use a promising new capability. As an example, AFOTEC was involved along with the Defense Evaluation Support Activity (DESA), at the request of the Deputy Under Secretary of Defense/Anti-Terrorism (DUSD/AT), in analysis and assessment of the Predator UAV when it was the subject of an ACTD and used in Bosnia.

Examples of Air Force Experimentation

The predominant thrust of the Air Force Experimentation Program has centered on concepts and initiatives to achieve an effective expeditionary Air Force. Under the management of the AFEO, a series of experiments (the JEFX-1998/1999/2000/2002) examined command and control and information sharing (including coalition and alliance partners) over global networks and the relationships between sensors and weapons as they apply to the time critical target (TCT) problem. These activities were coordinated with the USJFCOM Rapid Decisive Operations concept development to ensure that joint aspects were fully appreciated and accommodated. This work has resulted in significant improvements in TCT operations in Afghanistan, as compared with those of Desert Storm. Significant reductions were achieved in the time required for executing the "kill chain."[48]

---

[48]See Anthony H. Cordesman, 2002, *The Lessons of Afghanistan: War Fighting, Intelligence and Force Transformation* (Significant Issues Series, Vol. 24, No. 4), Center for Strategic and International Studies, Washington, D.C., November, p. 110.

Examples of results that are planned to transition into USAF CAOC Combat Capability as a result of JEFX-2002[49] include the Mapping Tool Kit, the Blue Force Tracker, the Combat Search and Rescue Module, the Space Tasking component in the Air Tasking Order, and three other, classified capabilities. The top initiatives were screened by the Air Force and successfully competed for set-aside funding for transitioning the results of experimentation to the field. However, some portions of all of the initiatives are being transitioned to the warfighter. These are small items, including a mix of new software and materiel, improvements to existing systems, TTPs, and training improvements. All of these items resulted from spiral processes, specifically from seven spirals. Several of these, including materiel and TTPs, have resulted in "leave behinds" for the field. For example, the Nellis CAOC benefited from the command and control systems left as residual assets. All of these initiatives were in the DOTMLPF approval process at the time this report was being prepared.

A second set of examples of results is provided by the USAF experiments with the Predator, Global Hawk, and UCAVs. While the Air Force was not much involved in the Predator ACTD (which was managed for DARPA by the U.S. Atlantic Command, which is now the U.S. Joint Forces Command, and the Navy program executive office), the Air Force was designated lead Service for this ACTD by the Vice Chief/Joint Chiefs of Staff in late 1995. As a result of its strong interest in the possibility of a replacement for the U2, the Air Force was heavily involved in the ACTD for Global Hawk and became the lead Service at the beginning. Both Predator and Global Hawk were participants in JEFX-1999 and also JEFX-2002, although Global Hawk was simulated for cost reasons. It is interesting to note that the firing of the Hellfire missile from Predator was demonstrated in 2001. Notional UCAVs have been simulated in experiments, such as JEFX-1999, but UCAVs are still in the development testing phase. All of these experimentation activities are managed by the AFEO and appear in campaign plans. Data obtained from these as well as other types of events are analyzed and archived by the AFEO.

The use of UAVs as combat vehicles is on the threshold of major transformation in joint warfighting. Both the Predator and the Global Hawk are in the Air Force procurement program and include programs of record. UCAVs are still in testing stages under a joint DARPA and Air Force development program but are not in the Air Force Program Objectives Memorandum.

Another example of Air force experimentation is that of a series of air command and control experiments under way with the USAF "Paul Revere" Boeing 707. These experiments are also managed by the AFEO under the Air Combat Command/C2 ISR Center (ACC/C2ISRC). The results of these experiments are

[49]JEFX-2002 included the Air Force segment of USJFCOM's Millennium Challenge '02 experiment.

relevant for mission crew positions and equipment, associated processes and procedures, and the requirements for communications/bandwidth. Favorable results will be spiraled into a Boeing 767, which will be equipped with an improved electronically scanned synthetic aperture radar sensor. This Boeing 767 will become the prototype/experimentation platform for the future USAF MC2A. The concept of operations places the MC2A in control of on-board and off-board sensors with a fleet of UCAVs complementing an expeditionary CAOC. The Air Force has already proposed funding for the expedited development of an MC2A.

Finally, the USAF battle laboratories continue to produce combat capability improvements. Many of these constitute initiatives that provide hardware and software with TTPs. Battle laboratory initiatives have all been the subject of at least laboratory-level tests before becoming part of AFEO-managed experiments. Data resulting from experiments are regularly entered into DOTMLPF formats. A list of each activity and its status is provided in a 2000 USAF report.[50] Table 3.2 provides insight into the nature of the initiatives that are generated by battle laboratories. Some of these initiatives were selected as subjects for various experimentation events managed by AFEO. As such, these activities account for only one part of the Air Force Experimentation Program. Many of these initiatives are similar to the limited technical assessments (LTAs) of the Marine Corps, discussed earlier in this chapter.

From Experiments to the Field

A summary discussion of the processes used by the USAF to resource and transition experimentation results follows.

- *Large, expensive systems and programs.* Large systems such as Predator and Global Hawk have their experimental genesis in DARPA and the Service R&D communities. As Service experiments and ACTDs with these two air vehicles proceeded, their military utility eventually became so compelling that they gained senior officer support (from the Chief of Staff of the Air Force and the Secretary of the Air Force), and secured funding in the USAF POM.[51] Moreover, the UCAV and the MC2A (both supported by the Chief of Staff of the Air Force and the Secretary of the Air Force) yielded experimental results that formed

[50] Lt Gen Robert H. Fogelsong, USAF, Deputy Chief of Staff, Air and Space Operations. 2000. *Air Force Experimentation Campaign Plan FY00-05,* Department of the Air Force, Langley Air Force Base, Va.

[51] In 1995, the Predator secured Air Force POM funding as a consequence of VC/JCS designation of the Air Force as the lead agency. In 1996, the Chief of Staff of the Air Force stated that UAVs would play a significant role in the future battlespace environment; this was a turnaround from Air Force policy of the previous 20 years. Conceivably it was this four-star endorsement that removed opposition and enabled POM funding.

the basis for their competing in the normal USAF POM process. Nonetheless, it is senior advocacy that provides the entrée into the competition.[52] Those efforts associated with programs that require significant funding have, as a general rule, little chance of being funded and transitioned into the force. Exceptions may occur if the efforts are pushed very hard by a very senior leader proponent (influential four star).

• *Less-expensive programs and systems ($50 million to about $70 million).* The Air Force battle laboratories and AFEO activity produce this class of candidates for transition almost without exception. To convert experimental systems to combat capability becomes a matter of obtaining the necessary resources through the POM process. The battle laboratories and the AFEO initiatives each have a slightly different transition process for obtaining their place in the POM for funding (discussed below).

• *Battle laboratory initiatives.* For the past 3 years, the USAF has established a source of funds called the Warfighter Rapid Acquisition Program (WRAP). The purpose of the fund is to bridge the gap in the POM and to sustain initiatives until they can compete in the POM process; at that point, if they are successful, POM funding takes over. Originally $30 million to $40 million was allocated in this fund, but it is common for this fund to support other priorities, and it is often reduced to about the $10 million level. In addition, unless sponsored by a very senior officer (four star), which seldom occurs, the worthy battle laboratory system initiatives are seldom successful in obtaining POM funding; this makes the decrease in WRAP funding a moot issue.

• *AFEO initiatives.* For the past 2 years, and instantiated with JEFX-'02, the AFEO has modified processes to transition its experimental results into improved C2 and ISR combat capability. Since it is meeting with success, the process is worth elaborating on here. Using JEFX-'02 as a case study and beginning 2 years in advance (FY 2000), the process starts by soliciting initiatives for experimentation. For the next 3 to 5 months, the initiatives are vetted first through the warfighting communities of the USAF and then to the Chief of Staff of the Air Force, who provides final approval for those initiatives selected for the JEFX experiment. This vetting process is very extensive, and by the time a system initiative is approved by the USAF, all of the underlying analysis and study have been accomplished and used to justify its selection.

Because of the rigor of the vetting process, by approval time there is little doubt that if the experimentation for testing the initiative is successful (using criteria established in the selection process), the initiative will be transitioned into

[52]There is significant support from the Chief of Staff of the Air Force for the MC2A. Gen John Jumper, as the Air Combat Commander, emphasized the experimentation that is providing critical data for the MC2A suite. He is now sponsoring MC2A as a major item in the R&D budget.

the force. During this period, and as part of the process, a corresponding transition plan is developed and approved (by the Chief of Staff of the Air Force) simultaneously with the selection of the initiative itself. A part of this transition plan is to identify the program element in the POM that will subsume a particular initiative if it is found worthy. In this case, the program element was the Air Operations Center Improvement, which has a considerable amount of funding. The goal would be, to the extent possible, to allocate or earmark funds for an initiative under consideration. Alternatively, if necessary, funds can be reprioritized and reprogrammed within the program element in the future, if and when the initiative meets experiment exit criteria.

In addition, transition or bridge funding is allocated from the JEFX funding to sustain the initiative as needed until the POM funding stream becomes available. About $10 million of JEFX-2002 money was so earmarked and employed.[53]

As noted earlier, seven major initiatives successfully met the predetermined exit criteria in JEFX-2002. The $10 million bridge funding was used for each. The results of the experiments were briefed to the Chief of Staff of the Air Force in late September 2002 with a recommendation that they be accommodated in the Air Operations Improvement program element in the FY 2004 POM. The Chief of Staff of the Air Force approved the recommendation, and the seven initiatives were made part of the FY 2004 POM, with bridge money to sustain them until FY 2004. Initial results suggest that this is a successful process for moving small(er) programs to the field.

Future Air Force Plans and Programs for Experimentation

The Air Force Experimentation Campaign is a rolling 6-year plan that is updated each year with designated thrust areas. Most assuredly the overall thrust toward expeditionary capabilities will remain, and for the AFEO experiments, the C4ISR emphasis. In addition to the ongoing experimentation work in the USAF battle laboratories, the Air Force will concentrate heavily on the many-faceted aspects of the future USAF Command and Control Constellation. This includes sensors, UAVs, processes, procedures, and its centerpiece aircraft, the MC2A, and includes coordination with the Global Strike Task Force using specific Service experiments or as part of joint experimentation. More effort will be spent on time factors and tracking in time-critical targets. The battle laboratories and the Air Force Experimentation Office, using the GSTF facility, plus infrastructure at Hurlburt Field, Nellis and Eglin Air Force Bases, and other places, will continue to work on the associated tactical innovations and experiments and their interfaces to strategic and operational work. Direct influence for innovations

[53]Per committee discussion with AFEO staff on September 9, 2002.

in strategies is anticipated from the experiences of expeditionary air forces now deployed.

Some methods and tools for experimentation will be enhanced in the future. The adequacy of simulation and data analyses will be assessed, since the AFEO believes that improvements are needed to raise confidence and increase depth. Also, a Joint Synthetic Battlespace will be explored. The Air Force has been applying spiral processes for some years, and a macrospiral process will be investigated—involving spirals that include changes in concepts of operation as well as in TTPs.

4

Emerging Roles in Experimentation—The Joint Connection

All recent U.S. military operations have been joint, and nearly all future operations will likely be so as well. Hence, naval concept development and experimentation should be done in a joint context, although there are naval-unique capabilities that must also be developed at the same time. One challenge for the Naval Services is to establish the proper balance between the joint and naval-specific activities.

This balance is by no means static. Indeed it has been undergoing profound shifts in the last few years as both the Department of the Navy and the entire Department of Defense have rethought and reworked their experimentation efforts. While most of the committee's data gathering drew to a close during the summer of 2002, it believes that fundamental lessons for experimentation can be discerned even as this evolution in experimentation continues in the Department of the Navy and the Department of Defense.

Overall, three distinct venues for joint experimentation present themselves today: within the U.S. Joint Forces Command (USJFCOM) experimentation process, within the regional Combatant Commands in the field, and through cross-Service experimentation, in which another Service participates in a predominantly naval experiment or vice versa. This chapter discusses each of these venues and provides a set of basic principles for striking an appropriate balance between joint and naval-specific experimentation.

U.S. JOINT FORCES COMMAND AND ITS EVOLVING MISSION

Today's U.S. Joint Forces Command was foreshadowed by the U.S. Readiness Command in the 1970s and 1980s, which then had the mission of developing joint doctrine and evaluating joint readiness through joint exercises and other means. First a regional command, USJFCOM has lost geographical responsibilities and is now a functional unified command, primarily a force provider, joint force trainer, with additional responsibilities in joint requirements and interoperability. Since Congress specified in 1998 that the Secretary of Defense should designate a combatant commander to undertake joint warfighting experiments and has since directed the DOD to undertake specific "Joint field experiments,"[1] USJFCOM now plays an important and growing but still changing role in military experimentation.

Historically, USJFCOM evolved from the United States Atlantic Command (USLANTCOM), which had long focused on defending the Atlantic Ocean and also on operations in the Caribbean.[2] In recognition of the changing world environment, the mission of USLANTCOM was changed in 1993 to provide a more joint focus (and its acronym changed to USACOM). In particular, component commands from the Marine Corps, Army, and Air Force were added to the Navy component command already assigned, and USACOM was made responsible for training forces from all four Services for joint operations. USACOM would then supply ready joint forces to other unified commands anywhere in the world, while still maintaining its responsibilities in the Atlantic and the Caribbean.

In 1999, the U.S. Joint Forces Command (USJFCOM) was created from USACOM. It retained the responsibilities and Service component commands of USACOM but was given even greater joint focus, with new responsibility for joint force integration, experimentation, and doctrine development in addition to joint training. Then, so that USJFCOM could focus entirely on these functional responsibilities and on its role as a force provider, its geographic areas of responsibility were assigned to other unified commands in 2002.

Today, USJFCOM figures prominently in the work of defense transformation, including the development of joint concepts, the conduct of experiments, and the definition of joint requirements. In USJFCOM's own statement of its vision, "U.S. Joint Forces Command leads the transformation of the United States

---

[1]That a combatant commander should be designated to undertake joint warfighting experiments is expressed as the sense of Congress in the National Defense Authorization Act for Fiscal Year 1999 (P.L. 105-261, 112 Stat, 1920. For an example of its directing the DOD to undertake specific joint field experiments, see National Defense Authorization Act for Fiscal Year 2001 (P.L. 106-398, 114 Stat. 1654).

[2]A good article on USJFCOM's history is that by ADM Harold W. Gehman, USN, (retired), 2000, "Progress Report on Joint Experimentation," *Joint Forces Quarterly*, No. 25, Summer, pp. 77-82.

Armed Forces to achieve full spectrum dominance as described in Joint Vision 2020."[3]

USJFCOM is the executive agent for joint warfighting experimentation. Its congressional mandate from the National Defense Authorization Act for Fiscal Year (FY) 2001 requires it to conduct "a Joint warfighting experimentation program" through "field experiments under realistic conditions across a full range of future challenges."[4] USJFCOM's most recent response to this mandate was a large field event named Millennium Challenge '02 (MC 02) discussed at length below. MC 02 was embedded in a larger experimentation context of war games, analyses, and limited-objective experiments (LOEs). As will be seen, this has proven a resource-intensive requirement in terms of people, operational units, and experimental infrastructure as well as equipment and funding.

USJFCOM's evolution appears by no means to be at an end. In fact, its pace of change may be increasing as recent guidance from the Chairman of the Joint Chiefs of Staff leads to further changes in the ways in which USJFCOM plans and conducts experiments. The most relevant portions of this latest guidance read as follows:

1. As Executive Agent for joint experimentation, you will develop a Joint Experimentation Campaign Plan (JE CPLAN) that looks both inside and outside the Department of Defense for concepts and capabilities. . . . The plan must incorporate a decentralized process to explore and advance emerging joint operational concepts, proposed operational architectures, experimentation and exercise activities currently being conducted by the Joint Warfighting Capabilities Assessment Strategic Topic Task Forces, the combatant commands, the Services and Defense agencies. . . .
2. The development of a standing joint force headquarters (SJF HQ) prototype and the other tasks directed and outlined . . . remain the highest priority.
3. In coordination with the combatant commands, Services, Joint Staff and Defense agencies, include the following in USJFCOM's JE CPLAN:
   a. Rapid exploitation of ideas and innovations demonstrated in, and lessons learned from, the war on terrorism.
   b. Concepts, capabilities and measures of effectiveness to conduct joint operations in an uncertain environment and complex terrain. Include for approval the concepts and capabilities for improvements in joint operation and command, control, communications, computers, intelligence, surveillance and reconnaissance (C4ISR) in urban terrain and jungle environments, and consider joint operations in mountainous or heavily forested environments. Apply special emphasis to the concepts in limited objective experiments and other events in FY 2004 and FY 2005.

---

[3]GEN Henry H. Shelton, USA, Chairman of the Joint Chiefs of Staff. 2000. *Joint Vision 2020*, The Pentagon, Washington, D.C., June, p. 13. Available online at <http://www.dtic.mil/jointvision/jv2020.doc>. Accessed October 7, 2003.

[4]National Defense Authorization Act for Fiscal Year 2001(P.L. 106-398, 114 Stat. 1654).

c. Fast-deploying joint command and control structures that . . . [employ] reach-back. . . .
d. Concepts to provide warfighters at all levels improved real-time battlespace awareness. . . .
e. Joint capabilities enabling the near-simultaneous . . . deployment of air, land, sea, cyber, and space warfighting capabilities. . . .
f. The transformational concepts and capabilities of the Nuclear Posture Review. . . .
g. Current efforts to promote and develop regional component commander-sponsored joint and multinational experimentation and capability-based modeling and simulation partnerships.

4. It is important to ensure continued development of the concepts and ideas demonstrated during and emerging from Millennium Challenge 02. . . .[5]

Roles and Capabilities of the U.S. Joint Forces Command

USJFCOM currently offers capabilities that support both experimentation and transformation. It will play an increasingly strong role in both in the future. Its main activities to these ends revolve around concept development, scheduling, experimentation, doctrine and training, and requirements definition, as briefly noted below.

- *Concept development.* USJFCOM's Joint Experimentation Directorate (J-9) organization is developing a number of new joint warfighting concepts with an emphasis on operational-level unified command and control. Many USJFCOM concepts in the knowledge-management area are broadly similar to those of the Navy Warfare Development Command (NWDC), but most are still early in the concept-generation process. As a concrete measure of their present maturity, none of the concepts had been validated or accepted at the time of this study, and only about half had received an initial assessment as of mid-2002 when MC 02 was carried out.
- *Scheduling.* USJFCOM conducts annual scheduling conferences that coordinate and synchronize deployments, exercises, and other major activities across the Services—including experimentation—in order to maximize the performance of the joint mission.
- *Experimentation.* USJFCOM conducts a range of experiments, including the congressionally mandated joint series of Millennium Challenge exercise/experiment. These joint "field experiments" are a combination of what this study

[5]Gen Richard B. Myers, USAF, Chairman, Joint Chiefs of Staff. 2002. "Guidance for USJFCOM Joint Experimentation [Enclosure]," Memorandum for the Commander, U.S. Joint Forces Command, November 26, p. 2.

terms exercises and experiments that explore USJFCOM concepts. At the same time, these experiments may also explore Service-specific concepts, through the incorporation of Service events such as fleet battle experiments (FBEs) into the bigger joint experiment. Each joint experiment includes a lengthy planning cycle involving all of the Services. The NWDC and the USMC's Joint Concept Development and Experimentation (JCDE) division work closely with the USJFCOM experimenters in the J-9 organization.

- *Doctrine and training.* USJFCOM, through its Joint Warfighting Center (JWC), develops joint doctrine and trains all joint headquarters staffs in it. The JWC includes the Joint Training, Analysis, and Simulation Center (JTASC) for performing the training. The JTASC also contains USJFCOM's Joint C4ISR Battle Center (JBC), chartered to lead the near-term transformation of USJFCOM's C4ISR capability through assessment and experimentation with new technologies. The JBC identifies shortcomings in the combatant commanders' C4ISR systems and provides near-term solutions. It also prepares DOTMLPF documentation to support the solutions.
- *Requirements.* USJFCOM's joint interoperability and integration (JI&I) process coordinates joint experimentation results to ensure that all DOTMLPF recommendations are integrated for consideration by the Chairman, Joint Chiefs of Staff (CJCS) and the Joint Requirements Oversight Council (JROC).[6] The JROC, composed of the vice chiefs of each Service under the vice chairman, appears to be gradually enlarging its mandate to include a role in the approval of joint concepts and perhaps experiments, in addition to its established role of approving requirements.

Current and Evolving U.S. Joint Forces Command Experimentation Programs

USJFCOM conducts many different types of experimentation activities, including war games, constructive simulations, human-in-the-loop simulations, limited objective experiments, and large-scale field exercises such as Millennium Challenge '02. Table 4.1 presents USJFCOM's experimentation activities through September 2002.

The active forces in these USJFCOM experiments, as well as large-end items such as equipment, headquarters, staffs, and so forth, are provided by the Services, which in general are performing their own Service-unique experiments simultaneously. The Services are directed to support, participate in, and fund a share of the joint experiment effort, over and above their own experiments.

---

[6]This is an important enough activity in practice that the Marine Corps recently signed a memorandum of agreement between the Marine Corps Combat Development Command and the JI&I element to ensure that the Marine Corps properly participates in this integration effort.

TABLE 4.1 Experimentation Activities of U.S. Joint Forces Command Experimentation, 1999–2002

Date of Experiment	Experiment Name	Type of Experiment
June/September 1999	Attack Operations 2015 Against Critical Mobile Targets	Human-in-the-loop virtual simulation
January 2000	Non-kinetic Technologies Limited Objective Experiment	Constructive simulation
June/August 2000	Rapid Decisive Operations Joint Analytic War Game	War game with constructive simulation
June/October 2000	Attack Operations Against Critical Mobile Targets	Constructive simulation
July/September 2000	Millennium Challenge 2000 Joint Field Experiment	Field experiment
May 2001	Information Presentation LOE	Laboratory staff experiment
June 2001	Rapid Decisive Operations in 2007 Joint Concept Refinement Experiment	War game with human-in-the-loop simulation
October/November 2001	Operational Net Assessment LOE	War game with human-in-the-loop simulation
December 2001	Effects Tasks Order LOE	War game with human-in-the-loop simulation
July/September 2002	Millennium Challenge 2002 Joint Field Experiment (MC 02)	Field experiment

SOURCE: Richard Kass, Chief, Analysis Division, U.S. Joint Forces Command, presentation to the committee on July 30, 2002, Norfolk, Va.

At present, major joint experiments are scheduled at 2-year intervals. These high-level experiments, exercises, and demonstrations require ongoing coordination, as well as interaction by all the Services to ensure that Service equities are protected in terms of cost, operational tempo, and forces participating. At the time of this writing, the first major joint experiment, Millennium Challenge '02, had just finished, so most of the lessons learned to date from this large-scale field exercise/experiment have come from its planning and preparation stages.

Millennium Challenge '02 (MC 02)

Millennium Challenge '02 was a congressionally mandated field exercise/experiment that combined a number of goals, including troop training and Service experiments (e.g., the Navy's Fleet Battle Experiment-Juliet (FBE-J) and the Marine Corps Millennium Dragon 2002 experiment). The mission statement for

MC 02 reads as follows: Conduct Millennium Challenge 2002 from July 14 to August 15, 2002, using the western training and testing ranges to (a) determine the extent to which the joint force is able to implement the principles of Joint Vision 2020 to execute rapid and decisive operations in this decade; and (b) produce [DOTMLPF] recommendations for [QDR 05 and budgets 04-09 on those actions that must be accomplished to] ensure the success of the Joint force in this type of operation.[7]

It is noted here that USJFCOM was not able to design, plan, and execute MC 02 in accordance with its desired plan. Although USJFCOM had been working on MC 02 for some time, the official "tasking" for the event was received via congressional mandate in the National Defense Authorization Act for 2001. In essence, USJFCOM was asked to develop and execute a major experiment in a very short time.[8] Given this guidance, USJFCOM conducted MC 02 as required.

By any measure, MC 02 was a large and complex field exercise. Approximately 13,500 personnel took part at 17 simulation locations and 9 live-force training sites, in a large, networked environment ranging from the seas off southern California, through the Midwest, and to a number of East Coast locations. Some 42 models and simulations were linked together, and to the live players, through networks to form a one-time Joint Experimentation Federation. The exercise involved a Navy battle group, a Marine expeditionary brigade, Army airborne and medium brigades, an aerospace expeditionary force, and joint special operations task forces. As mandated, the exercise itself lasted approximately 3 weeks, of which 10 days were devoted to live action.

In a briefing to the committee, USJFCOM provided observations based on lessons learned in the workups to MC 02. The observations refer to the capabilities explored in MC 02 (see Box 4.1). These points formed the basis for discussion in the after-action review that immediately followed MC 02.

A summary of the Navy's assessment of its participation in MC 02 is given later in this chapter.

The Costs of Millennium Challenge '02

Several times the committee heard that the total cost of MC 02 was approximately $250 million, including the costs of the Services and of USJFCOM.

---

[7]CAPT Richard A. Feckler, USN, U.S. Joint Forces Command, J9239, "The Millennium Challenge, 'The Scene Setter,'" presentation to the committee on July 30, 2002.

[8]The "normal" time is the 6-year cycle shown in Figures 4.2 and 4.3 in this chapter. The figures are from USJFCOM's planning documents. As USJFCOM got off to a late start for MC 02 (owing primarily to the first-time nature of many aspects of MC 02 and the congressionally mandated dates for the experiment), it had about 3 or 4 years to plan and execute MC 02. As the experimentation cycle evolves, there will probably be as many as three current experiments in the cycle at any one time.

BOX 4.1
Capabilities Explored in Millenium Challenge '02

- *Effect-based operations.* A new way to think about mission accomplishment; which has the potential to provide an advantage to all elements of national power to influence or deter an adversary.
- *Standing joint force headquarters.* Reduces the ad hoc nature of current JTF [Joint Task Force] operations and can reduce the planning times dramatically and improve interoperability.
- *Interagency community.* Early and continuous interagency involvement in a crisis.
- *Operational net assessment.* Comprehensive knowledge of adversaries' and allies' capabilities and of the nature of the battlespace.[1]

[1]CAPT Richard A. Feckler, USN, U.S. Joint Forces Command, J9239, "The Millennium Challenge, 'The Scene Setter,'" presentation to the committee on July 30, 2002.

However, it has not been able to verify this figure or to determine which costs might be included in this total. It is true that significant forces of each of the Services were fielded; these included major elements at regiment, wing, and battle group levels. However, since the Services also were conducting their own experiments (such as FBE-J) at the same time, the incremental "joint" cost of MC 02 was almost certainly lower than this total. This amount may not be excessive from the standpoint of meeting the congressional requirement for joint field experiments. However, the committee believes that to apply resources effectively, it is necessary to select those venues that may be considerably less expensive but that provide the required knowledge (e.g., wargaming, simulations, and smaller, more focused live experiments) and reserve the larger-scale field experiments for objectives that only they can address.

U.S. Joint Forces Command's Emerging Concepts

Table 4.2 lists the current set of joint concepts that USJFCOM is formulating together with the committee's understanding of its current assessment status. To date none of the concepts has been validated or accepted, but some have been subject to initial assessment in MC 02. It is still too early to comment further on these concepts.

TABLE 4.2 Emerging Concepts from the U.S. Joint Forces Command

Name of Concept	USJFCOM Description	Subject to Initial Assessment in MC 02?
Effects-Based Operations (EBO)	EBO is defined as a process for having a desired strategic effect on the enemy through the synergistic and cumulative application of the full range of military and nonmilitary capabilities at the tactical, operational, and strategic levels.	Yes
Rapid Decisive Operations (RDO)	RDO is an evolving concept for how the joint force, acting within an interagency context, can defeat a capable regional power in weeks rather than months. It explores future joint operations with future military and crisis management.	Yes
Standing Joint Force Headquarters (SJFHQ)	USJFCOM postulates an SJFHQ assigned to each theater combatant commander and embedded in the combatant commander's staff under the direction of a flag or general officer. When a contingency requires the establishment of a Joint Task Force (JTF), the SJFHQ can immediately form the JTF.	Yes
Adaptive Joint Command and Control (AJC2)	AJC2 is focused on how the SJFHQ internal organization and joint force information transfer and C2 processes can best be developed to ensure the full realization of the forces' capabilities to conduct RDO and to adapt rapidly to changing information.	No
Operational Net Assessment	Operational net assessment envisions a systematic review of enemy motivations, objectives, alternative courses of action, and sources of strength and weakness, together with an equivalent understanding of U.S. and allied strengths and vulnerabilities as they bear on a conflict.	Yes
Joint Interactive Planning	Joint interactive planning is a parallel, collaborative, and adaptive planning process enabled by distributed interactive information systems that will allow supporting staffs and centers of technical expertise, separated by geography, time, and organizational boundaries, to interact in developing options and plans.	Yes
Interagency and Multinational Operations	This new J-X staff element would include members from other U.S. departments as well as from coalition states and possibly nongovernmental organizations. The J-X would facilitate interagency coordination and effective multinational operations.	Yes

continues

TABLE 4.2 Continued

Name of Concept	USJFCOM Description	Subject to Initial Assessment in MC 02?
Common Relevant Operating Picture (CROP)	The CROP will store and make immediately available current and archived information as appropriate to all force and command echelons, including (through the use of multilevel security) multinational partners and nongovernmental organizations.	No
Assured Access	Assured access refers to the ability to rapidly set and sustain the battlespace conditions necessary to bring the joint force within operational reach so that it can have the desired effects on an adversary.	No
Focused Logistics	Focused logistics addresses capabilities necessary to achieve rapid force deployment and agile sustainment, and to help assure access.	No
Joint Intelligence Surveillance and Reconnaissance (JISR)	JSIR is an actively managed joint capability directing sensors at all levels to collect and integrate information of the highest strategic, operational, and tactical value.	No
Information Operations (IO)	Information operations are the elements of information superiority that focus on the attitudes and perceptions of decision makers and the information systems and processes that support decision making. As such, IO is the information equivalent of maneuver and fire power.	No

SOURCE: The committee derived this overall list of concepts and descriptions from USJFCOM's document *The Joint Concept Development and Experimentation Campaign Plan, FY 2002-2007,* dated Feb. 11, 2002, pp. 15-18; the list of concepts included in MC 02 from USJFCOM's pamphlet entitled "Millennium Challenge 2002 Experimental Objectives"; and initial assessment results from the Navy's "Quicklook Report on Fleet Battle Experiment Juliet," dated August 2, 2002.

Synopsis of Results of U.S. Joint Forces Command Experimentation to Date

The results of joint experimentation should, in coming years, flow directly from USJFCOM's mission to "(a) discover promising alternatives through joint concept development and experimentation; (b) define enhancements to joint warfighting requirements; (c) develop joint warfighting capabilities through joint training and solutions; and (d) deliver joint forces and capabilities to warfighting commanders."[9] Given this mission, joint experimentation results will consist mainly of DOTMLPF recommendations and warfighting concepts. However, since joint experimentation is still in its early stages, as yet it has resulted in few changes.

As of this writing, USJFCOM had submitted for action three sets of DOTMLPF recommendations to the Joint Staff Force Structure, Resources, and Assessment Directorate (J-8) for presentation to the JROC, but none had yet been approved. These recommendations concern collaborative environments and tools, training for time-critical targeting related to theater missile defense, and joint intelligence preparation of the battlespace, again focused on theater missile defense.

The recent efforts of USJFCOM in Millennium Challenge '02 and for the next several years focus on two overarching concepts: (1) improving joint command and control through Standing Joint Force Headquarters (SJFHQ) and (2) conducting more effective joint operations through a new approach called Rapid Decisive Operations (RDO), although this latter focus may shift to the Joint Capstone Concept being developed by the Joint Staff.

The Chairman of the Joint Chiefs of Staff and the Secretary of Defense have directed USJFCOM to continue prototyping and validation of the SJFHQ in order to deliver its software and hardware components to each of the regional combatant commanders. No JROC decision will be needed in this case. Prototyping and experimentation will be carried out primarily through Combatant Command exercises. The first installations of these prototypes are currently planned for FY 2005 (at the end of calendar year 2004). The USJFCOM Joint Training and Joint Warfare Center (J-7) has the lead on actual delivery, but SJFHQ activities are also currently the top priority for J-9, which will be heavily involved in further conceptual development and experimentation.[10]

RDO is still in the experimental stage, but the first experiments with its associated concepts appear promising to USJFCOM. As this report is being written, it appears that USJFCOM's focus is shifting toward the Joint Capstone Concept now in review by the Joint Staff. In addition, J-9 has written a Joint Concept of Operations, and the Army and Air Force have produced a Joint Operational Concept. All of these concepts have been informed by the basic tenets of the RDO concept. These three overarching concepts, and possibly others, will be examined and competed in a series of war games in the campaign "Pinnacle Impact," and the result will form the basis for USJFCOM's future concept development activities.[11]

---

[9]USJFCOM mission statement, available online at <http://www.USJFCOM.mil/about/about1.htm>. Accessed December 1, 2002.

[10]According to committee correspondence with USJFCOM J-9 support staff, August 2002.

[11]According to committee correspondence with USJFCOM J-9 support staff, August 2002.

Future Experimentation of the U.S. Joint Forces Command

USJFCOM's future plans involve a 6-year integration of concept development, experimentation, and prototyping through analysis, gaming, and several levels of both joint and Service experimentation (Figure 4.1).[12] This plan defines a concept development path, which includes the new "Title 10 Joint Wargame,"[13] also called Joint Global Wargame (JGW). It also defines a "continuous experimentation environment" path, with spirals of workshops, games, LOEs, red teams, and so on. In parallel with these are Millennium Challenge, Olympic Challenge (OC), and other major events including FBEs, Joint Expeditionary Force Experiments (JEFXs), and AWEs.

This strategy puts USJFCOM at the center of all concept development and experimentation. Moreover, USJFCOM conducts the training of the Combatant Commands' Joint Headquarters. USJFCOM is in a position to exercise considerable influence on all Service experimentation, and its emerging Continuous Experimentation Environment may significantly affect future experimentation. In short, as USJFCOM's role in experimentation grows, its campaign plans are likely to significantly influence naval experimentation in the near-term future. Consequently, NWDC has established a liaison element at USJFCOM, and the Marine Corps has provided an on-scene detachment of its Joint Concept Development and Experimentation (JCDE) division.

USJFCOM's first track, the concept development track, comprises Service Title 10 war games in the odd-numbered years, alternating with Joint Global Wargames in even-numbered years. JGWs are intended to be "capstone" joint war games, similar in design and construct to Service Title 10 war games, which serve as the integrating vehicle for Service concepts and capabilities. JGW 2004 will be centered on a future joint warfighting concept with associated enabling capabilities, with emphasis on the next decade's application of Effects-Based Operations. A number of earlier workshops, white papers, and experiments will lead up to this event, including Multinational LOE No. 2, Army Transformation Wargame 2003, Pinnacle Impact 2003, Navy Global Wargame 2003, and so forth.

The second track is designed to prototype, refine, and validate the first path. It currently consists of joint field exercises (Millennium Challenge, Olympic Challenge) in the even-numbered years, with Service experiments such as FBEs, JFEXs, AWEs, and Marine Corps experiments wrapped into these joint experiments.

---

[12]Material in this section is drawn from the *Joint Concept Development and Experimentation Campaign Plan, FY 2002-2007*, prepared by Gen William F. Kernan, USAF, Commander in Chief, U.S. Joint Forces Command, Norfolk, Va., dated Feb. 11, 2002, and from e-mail correspondence of the committee with USJFCOM staff in August 2002.

[13]Gen William F. Kernan, USAF, Commander in Chief, U.S. Joint Forces Command. 2002. *Joint Concept Development and Experimentation Campaign Plan, FY 2002-2007,* Norfolk, Va., February 11.

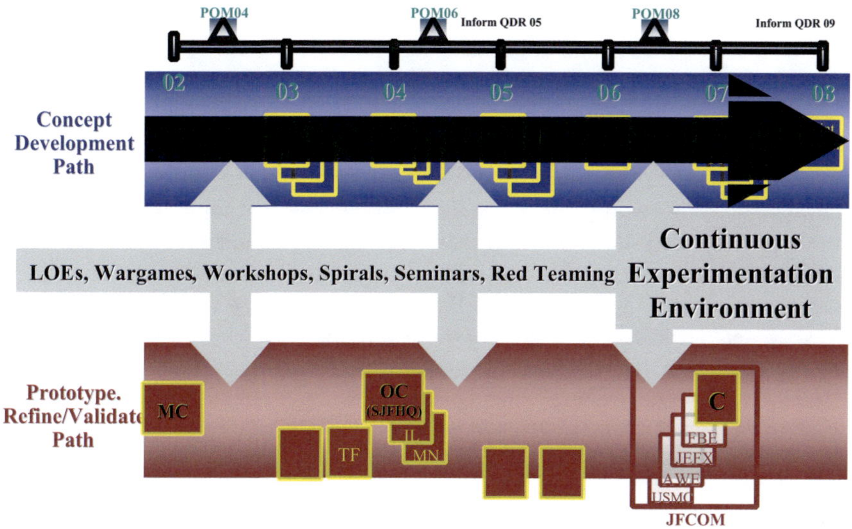

FIGURE 4.1 Emerging transformation strategy of the U.S. Joint Forces Command. SOURCE: John A. Klevecz, Assistant Director for Futures Alliance, Joint Experimentation, U.S. Joint Forces Command, "Future Joint Warfighting," presentation to the committee on July 30, 2002, slide 18.

Several USJFCOM briefers expressed the opinion that Olympic Challenge 2004 will likely be smaller and less ambitious than MC 02, although this remains to be seen. In particular, fresh mandates could well produce another field experiment similar in scale to that of MC 02. If so, this would require USJFCOM to plan and execute a major experiment in one-third of the proposed 2004 timeframe.

Emerging Strategy of the U.S. Joint Forces Command for Experimentation Campaigns

On the basis of experience gained in recent years and in response to lessons learned in the Millennium Challenge process, USJFCOM has defined a regular schedule for annual experimentation events—the simulation-based Vision experiments, cycled with the field-based Challenge experiments. "Vision" events are focused on concept development. "Challenge" events are aimed at testing concepts more robustly in integrated field experiments.[14] The overarching opera-

[14]Material in this section is drawn from the *Joint Concept Development and Experimentation Campaign Plan, FY 2002-2007*, prepared by Gen William F. Kernan, USAF, Commander in Chief, U.S. Joint Forces Command, Norfolk, Va., dated Feb. 11, 2002. The details presented here could change with the issuance of the new *Joint Experimentation Campaign Plan* now under development.

tional concept or theme to be considered (such as RDO in this decade for MC 02) can be identified early and developed over time. The resources to support these experiments can also be planned in advance.

A regular sequence for these major events and the necessary preparation provide a "battle rhythm" for joint experimentation. As shown in Figure 4.2, this 6-year cycle integrates concept development, experimentation, and prototyping through analysis gaming, and several tests of Service and joint experimentation. It starts with workshops and seminars early in the campaign, war games and limited-objective experiments in midcampaign, and a large simulation experiment to culminate the campaign, followed a year later by a corresponding field experiment. The recurring nature of these activities is important to USJFCOM's ability to engage Service and Combatant Command staffs on a routine basis. Because of the desired close integration with Service and Combatant Command experimentation activities, understanding the rhythm of this continuous experi-

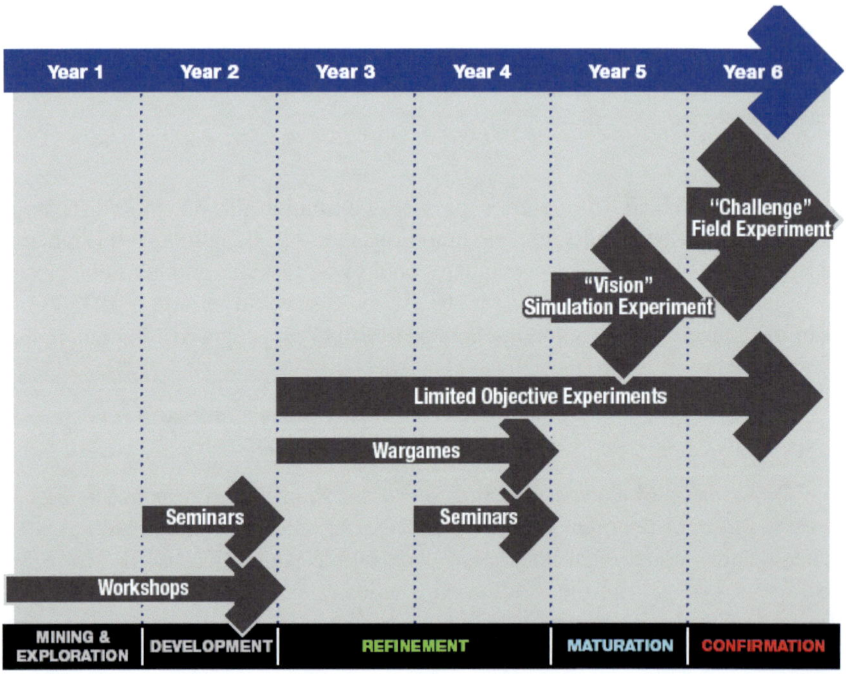

FIGURE 4.2 The U.S. Joint Forces Command 6-year schedule for an experimentation campaign. SOURCE: Commander in Chief, U.S. Joint Forces Command. 2002. "Figure 3.2: 6 Year Cycle of Concept Development and Experimentation," *The Joint Concept Development and Experimentation Campaign Plan FY2002-2007,* Norfolk, Va., p. 27.

mentation is important not only for internal USJFCOM planning purposes but also for experimentation activities throughout DOD.

Figure 4.3 superimposes the activities referred to above on the schedule for the USJFCOM series of experimentation campaigns for Olympic Challenge 2004, Pinnacle Challenge 2008, and Zenith Vision-09. These detailed, comprehensive schedules are required to support proper experiment design, planning, execution, and analysis, and the validation of results. Each campaign follows a 6-year pattern. At any given time, two different campaigns will be in progress, overlapped at different stages of the campaign development.

This 6-year cycle appears reasonable, considering the expenditure of resources required to execute a major experiment and the future potential value of the experiments' validated results. However, the committee notes that while this wide range of preparatory meetings and ramp-up events is in fact crucial for a well-run and useful joint experimentation campaign, it will surely impose a substantial burden on the Navy—and on the Marine Corps—to properly staff the USJFCOM experimentation campaigns. USJFCOM time lines will also surely influence those for naval experimentation. (These points are the subject of additional discussion in the section below entitled "Naval Linkages to U.S. Joint Forces Command Experimentation.") The committee also notes the need for and importance of maximizing the use of smaller venues, such as war games, simulations, and smaller field experiments and the continuation of activities beyond the culminating major event to derive the requisite knowledge.

Olympic Challenge 2004 (OC 04), to be conducted in the summer and early fall of 2004, will examine the capabilities needed by a Standing Joint Force Headquarters and the joint force to conduct Rapid Decisive Operations in the 2010–2020 timeframe against a high-intensity regional threat. The experiment is intended to describe "how we fight" as the joint force and to provide significant insights leading to DOTMLPF recommendations for changes to the 2005 Quadriennial Defense Review. OC 04 will have implications for choices of future military capabilities as well as the joint command and control (C2) enhancements that are the most likely focus of recommendations following MC 02. A period of intensive, smaller-scale experimentation and assessment across the joint community and the Services will be necessary after OC 04 to prepare high-confidence DOTMLPF recommendations.

OC 04 will build on and significantly extend MC 02. Following MC 02, USJFCOM plans LOEs to refine concepts for the SJFHQ and RDO, as well as supporting concepts, and the Services will refine their proposed capabilities. Concurrent with MC 02 and OC 04, a series of seminars and LOEs will lay the groundwork for the Olympic path joint concept development and experimentation (JCD&E). In 2003, USJFCOM intends to lead the joint community in conducting Olympic Vision 2003, a simulation-based experiment that will be the precursor to Olympic Challenge 2004.

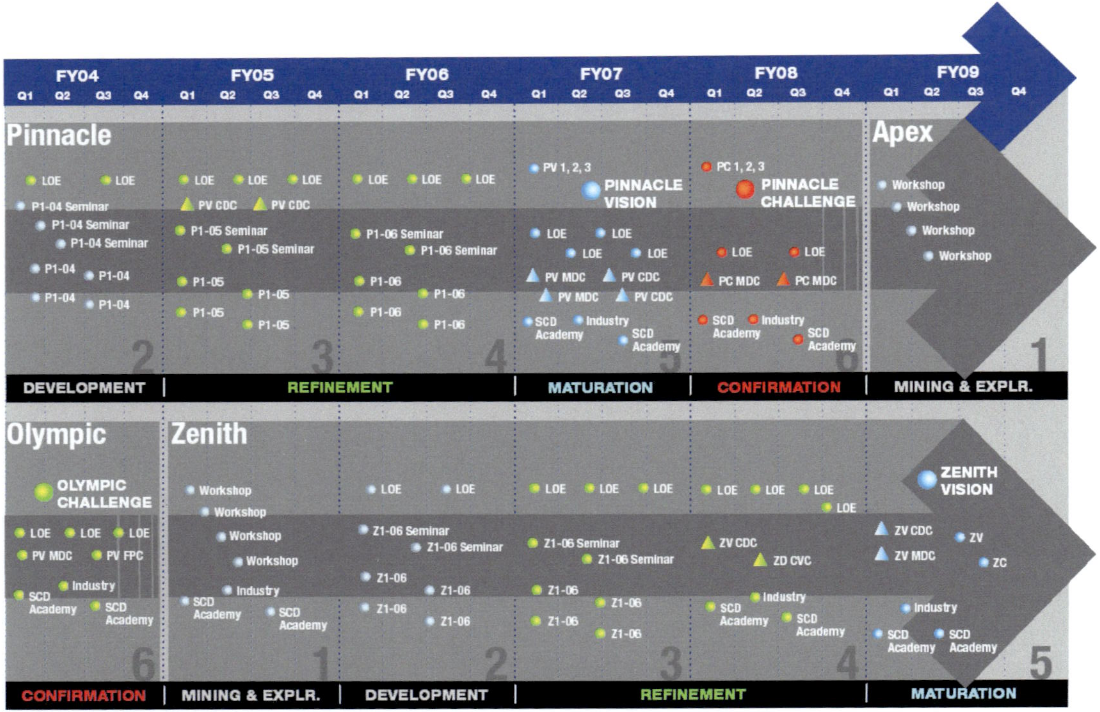

FIGURE 4.3 U.S. Joint Forces Command experimentation campaigns (Olympic, Pinnacle, Zenith). SOURCE: Commander in Chief, U.S. Joint Forces Command. 2002. "Appendix Figure on POM Planning Supporting CPLAN02," *The Joint Concept Development and Experimentation Campaign Plan FY2002-2007,* Norfolk, Va., p. 49.

Unlike MC 02, OC 04 will examine potential joint forces and capabilities that are not yet fielded and will be on a scale comparable to Unified Vision 2001. Through the use of prototypes, surrogates, and simulation tools—to the extent possible, all integrated by live joint headquarters with state-of-the-art planning and collaboration tools—USJFCOM will gain insight into the possibilities for major changes in future operations. The intent is that the Services, Combatant Commands, and others will bring their future force concepts and surrogates for their transformed capabilities to the experiment, which will evaluate advances in the tactical and operational levels of warfare, including integrated joint tactical actions. USJFCOM intends for the OC 04 scenarios to include actions by NATO, key allies, and coalition partners and envisions participation by non-U.S. forces in the experiment.

Later, Pinnacle Challenge 2007, to be conducted in the spring and summer of 2007, will explore the extent to which the principles of RDO can be applied to a broader range of the spectrum of joint operations as driven by key themes of a future security environment. This experiment will build on lessons learned from MC 02 and OC 04, as well as from a large number of smaller experiments conducted over the next several years.

U.S. Joint Forces Command's Emerging Modeling and Simulation Infrastructure

Because modeling and simulation can allow the rapid exploration of the capabilities of future technologies, organizations, and operational concepts without creating new hardware or consuming significant resources in terms of training or personnel, it can both accelerate the pace of experimentation and reduce its cost. Accordingly, it plays a large role in USJFCOM's vision of future joint experimentation.

In USJFCOM's view, the 42-model Joint Experimental Federation (JEF) satisfied the essential modeling and simulation needs of MC 02, but even so, this effort was only a first step. Many of DOD's modeling and simulation tools are based on a limited and legacy set of operational concepts, types of military effects, and defeat mechanisms, and thus are not useful for assessing the value of new operational concepts and capabilities.[15] Such legacy simulations are generally not capable of representing a nonlinear battlespace or one filled with a variety of operating units. Nor can they adequately account for information operations, deal with operations other than high-intensity conflict, model the efforts of

[15]In fact, the Navy's "QuickLook" assessment stated that the federation "created serious limitations in the execution of advanced concepts." See Commander, Navy Warfare Development Command, 2002, "Naval Message to Commander, Fleet Forces Command, Norfolk, Va., Regarding QuickLook Report Fleet Battle Experiment Juliet," NWDC, Newport, R.I., August.

nongovernmental organizations, or capture the effects of asymmetric warfare strategies. Many legacy simulations also lack the flexibility to integrate new elements on short notice at reasonable cost. Beyond the simulations themselves, the data analysis tools available to USJFCOM and the larger community are very limited. If the large amounts of data collected in simulations are to be examined in any substantive way, more sophisticated tools will be necessary.

In response to these perceived deficiencies, USJFCOM has identified a path to develop a continuous experimentation environment (CEE) to support all aspects of JCD&E, enabling more flexible, higher-fidelity modeling and simulations, more rapid iterations, and more confident recommendations. In this vision, the CEE would allow experimenters to perform near-continuous experimentation, both locally and at distributed locations to explore multiple questions simultaneously and to ensure that joint considerations are included in all experiments. A continuous experimentation environment would dramatically improve the DOD's capability to conduct joint concept development and experimentation.

In USJFCOM's view, the continuous experimentation environment CEE should employ models and simulations that recently formed parts of the JEF in whatever combinations are needed. For example, the Joint Semi-Automated Forces (JSAF) simulation, which was one of the core simulations for MC 02, would serve as a primary tool for entity-level combat. JSAF today can simulate up to 50,000 systems in a common battlespace and is flexible enough to support novel operational concepts and diverse friendly and enemy forces. However, USJFCOM estimates that a system 20 times as powerful would be needed to fully support the needs of joint experimentation. It is claimed that a scalable parallel processing system that can meet this requirement has been demonstrated in prototype form and that with moderate investment it could be available in a near-term timeframe. Service-developed tools would augment JSAF's high-fidelity simulation, and other JEF models would be federated as necessary to round out the environment.

Finally, USJFCOM is working to adopt for experimentation purposes the Joint Simulation System (JSIMS) training tool. JSIMS-E is intended to provide realistic, large-scale simulations placing humans in realistic operational environments; these simulations would allow interaction with other JSIMS participants and with sensor and other information provided either live or from an archive of earlier experimentation.

The implications of these USJFCOM modeling and simulation developments for the Navy and the Marine Corps are discussed below in the subsection entitled "Naval and Joint Linkages in Simulation."

The Role of Joint Experimentation in Preparing for Joint Operations

The whole point of joint experimentation is better joint operations. Today, each Service brings its own core warfighting capabilities to the fight. Examples include carrier-based air capabilities for the Navy, heavy divisions for the Army, stealth bombers and fighters for the Air Force, and amphibious forces for the Marine Corps. These warfighting capabilities are provided from the Services (supporting) to the Combatant Command (supported) for prosecution of the war campaign. But regardless of the command arrangement and structure, the Service forces fight as joint warfighters.

To date, joint and cross-Service experimentation has played three general roles in improving joint operations. Those roles can be summarized as follows:

- *Development of DOTMLPF for the Services and joint community.* These developmental improvements are enhancing Service core capabilities, joint capabilities, or both Service and joint capabilities.
- *Better understanding of other Services' capabilities.* Joint experimentation provides a directed forum for the joint participation by all of the Services in a common venue. The Services work together in designing and providing resources for the joint experiment. During these experiments, the coordination and interoperability of Service and joint capabilities (similar to the kinds of interactions required during joint operations) are explored and developed, all leading to improved joint operations.[16] Cross-Service experimentation programs provide a similar opportunity since they have reciprocal liaison staff and unit participation.[17]
- *The less tangible but important aspect of socialization.* Participants from different organizations with different Service cultures and perspectives work together toward focused goals. These efforts develop a better understanding and appreciation of the parts that make up the whole, resulting in both a better Service warfighter and a better joint warfighter.

The migration of new joint concepts into actual operations is not easy to trace. However, the committee heard specific anecdotal examples of successes applied in Operation Enduring Freedom. Major General James Mattis, USMC, Commanding General, First Marine Division (REIN) (then Commanding General, First Marine Expeditionary Brigade and the Joint Task Force Commander during portions of Operation Enduring Freedom), stated that some of the innovative operational concepts that he applied originated from his observations in the Hunter Warrior AWE. Rear Admiral David P. Polatty III, USN, Joint Task Force Commander, used several organizational concepts for his staff that resulted from a

[16]For example, through these experiments, Army forward air controllers are using a Marine Corps prototype, Precision Targeting System, while controlling Air Force and naval aircraft.

[17]For instance, Army, Air Force, and Navy officers are commonly assigned to the Marine Corps Warfighting Laboratory, and vice versa, for cross-Service experimentation efforts. This arrangement greatly facilitates interoperability of TTPs, especially in the areas of combined arms operations.

series of research experiments and a global war game. Some of these concepts were derived in part from the USJFCOM concept development and experimentation cycle in preparation for MC 02.

There is also evidence that operational experience is helping to guide experimentation plans. USJFCOM briefers stated that lessons learned from Operation Enduring Freedom in Afghanistan have been incorporated in their concepts for experimentation.

NAVAL LINKAGES TO U.S. JOINT FORCES COMMAND EXPERIMENTATION

Since naval operations are joint now and will most likely become even more so in the years ahead, linkages between naval and USJFCOM-sponsored experiments, and indeed the full range of experimental campaigns, should be carefully considered. The range of these campaigns runs from the earliest concept development through analysis, war games, and simulations, leading ultimately to LOEs and the large field experiments such as the Millennium Challenge series. An orderly progression of activities from concept development through experimentation will greatly aid in the efficient use of Service and joint resources.

In all phases of the campaigns, joint experimentation must coexist with Service-specific experimentation, because proficient Service core capabilities are a prerequisite for joint warfighting. Just as there is a specific hierarchy for training purposes (individual, unit, organizational, combined/joint), so there is a natural hierarchy for experimentation. The individual Services need ramp-up time before they bring new warfighting capabilities to the joint effort. To fine-tune their current capabilities and develop new capabilities, the Services must invent or improve their warfighting concepts, design experimentation programs, and conduct Service-unique experimentation to help validate their concepts. For major changes this can be a long, arduous process, requiring continuity and resources throughout a long-term experimentation campaign.

As described above, USJFCOM has laid out a detailed and intense joint experimentation campaign. The Services are directed to support and participate in this effort. These staff demands will stretch the already resource-limited Service-specific experimentation efforts. If, in the worst case, USJFCOM's activities interfered with or prevented the "normal" progression of Service experimentation, they could actually be counterproductive to developing improved joint capabilities. Thus, a balance must be maintained, and events and objectives must be synchronized between joint and Service-specific experimentation campaigns. In particular, it is important that Service experimentation programs progress in an orderly fashion so that they can improve the Services and joint capabilities.

Naval and Joint Linkages in Concept Development

As new warfighting concepts are developed, coordination among the Services and the joint community must ensure that individual concepts are compatible and mutually supportive. While concept development occurs throughout the Navy and the Marine Corps, it is focused for these Services, respectively, in the Navy Warfare Development Command and the Marine Corps Combat Development Command as discussed in Chapter 3. On the joint side, concept development is done in the U.S. Joint Forces Command and, more recently, also as part of the Joint Warfighting Capabilities Assessment (JWCA) process of the Joint Staff, the results of which are reported to the Joint Requirements Oversight Council.

The NWDC and the MCCDC interact with USJFCOM and (along with other elements of the Navy) the JWCA process in concept development activities, but in the committee's opinion this interaction is not particularly close. On the basis of information presented to it, the committee makes the following observations on this interaction:

- The extent of specific interactions with the joint community or other Services in the developing and refining of naval concepts does not appear to be large.
- The development of joint concepts by USJFCOM and the JWCA process does not appear to build on the naval concepts that are being developed by NWDC and MCCDC.
- The "joint community" itself does not at present speak with a single voice.[18] While responding to the same general themes, USJFCOM and the JWCA process are developing different sets of concepts.[19] A detailed correlation between these sets of concepts does not appear to have been established.

Naval and Joint Linkages in Simulation

"Simulation" in the present context refers to virtual and constructive simulations, as distinct from live forces. In this context, there are two subjects to be addressed: (1) naval participation in joint simulations and (2) the potential development and use of a common simulation environment by the Navy, Marine Corps, and USJFCOM.

Naval participation in joint simulations, which has occurred for years, is clearly valuable. At times the Navy participated in joint experiments involving

---

[18]As this study was being finalized, the Joint Staff was drafting the Joint Capstone Concept and the Joint Capabilities Integration and Development System document (Chairman of the Joint Chiefs of Staff Instruction (CJCSI) 3170.01C). Completion of the concept and its adoption by USJFCOM and realization of the processes specified in CJCSI 3170.01C could do away with the differences cited here.

[19]USJFCOM is developing such concepts as Effects-Based Operations, Rapid Decisive Operations, and Operational Net Assessment. The JWCA process is developing concepts from the Chairman's Joint Vision 2020 (e.g., Dominant Maneuver and Precision Engagement).

completely simulated environments (e.g., Unified Vision 2000 and 2001); however, it has primarily employed such simulations within the larger context of live field experiments that also involve actual deployed forces. It is true that deployed forces add a valuable element of realism, but much can be learned using virtual and constructive experiments, especially at the operational level of war. As the Navy begins, for example, to work through the concepts presented in the CNO document "Sea Power 21,"[20] virtual and constructive simulations can provide a very useful environment for developing, exploring, and testing the concepts of operation before trying them out with actual forces. Successful realization of these new concepts of operation may require many iterations within the next few years. This would be feasible in a simulation environment, but not in one requiring the use of deployed forces.

A series of (virtual and constructive) simulation-based limited-objective experiments will begin in FY 2003 to prepare for Olympic Challenge in FY 2004. Since the Navy will participate in these LOEs, it has an opportunity to better develop its use of virtual and constructive environments for experimentation purposes.

The development and use of a common naval/joint simulation environment are desirable, although the Navy and Marine Corps will also need their own naval-specific simulations. Such an interoperable environment brings twofold benefits: a consistent environment for both joint and naval-only play, and the cost economies of developing one environment to serve two (or more) communities. In the specific naval/joint context, a common environment should be greatly facilitated by the fact that the JSAF simulation forms the core of both the current USJFCOM and Navy simulation environments. The Marine Corps also uses JSAF,[21] although the Joint Conflict and Tactical Simulation (JCATS) was the basic simulation it used in Millennium Challenge '02.

USJFCOM has recently identified a significant list of desired enhancements to its modeling and simulation capabilities (see above, the subsection entitled "U.S. Joint Forces Command's Emerging Modeling and Simulation Infrastructure"). These enhancements could meet needs of the Navy and Marine Corps. Navy sources have indicated particular interest in greater fidelity in representing intelligence, surveillance, and reconnaissance data and weapons effects. Marine Corps sources indicated the need for greater human intelligence play (e.g., reconnaissance patrols). Another potentially useful area involves the representation of nonkinetic aspects of warfare—for example, information operations. This issue

---

[20]ADM Vernon Clark, USN. 2002. "Sea Power 21: Projecting Decisive Joint Capabilities," *U.S. Naval Institute Proceedings*, Vol. 128, No. 10, October, pp. 32-41.

[21]JSAF is used in the Marine Corps Distributed Virtual Training Environment (DVTE). In terms of the future evolution of DVTE, the Marine Corps could prefer to use OneSAF, developed by the Army, as the semiautomated forces component rather than JSAF, because of the greater configuration control imposed in OneSAF.

raises the overarching questions of how much of information operations is actually modeled in the simulation, and how much the simulation results depend on how the simulation is used (i.e., the operator actions driving the simulation). At the very least, it would seem necessary to have the simulation represent the capability degradation effects that can be brought about by information operations. Overall, USJFCOM and the Navy and Marine Corps (and other Services if they are involved) will need to agree on the priorities given to the various enhancements because their full undertaking would represent a significant expenditure of resources.

Naval Participation in Joint Field Experiments

The Navy and the Marine Corps have actively participated in joint field experiments for some years, with increased emphasis on these activities in recent events such as Millennium Challenge in the years 2000 and 2002. Representatives from both Services indicated to the committee the clear value of joint field experiments, but at the same time they noted the need for Service-specific experimentation so that the Services could adequately develop their own core competencies. They further pointed out the increased demands that joint experimentation places on their already highly committed forces and personnel, as well as the increased financial commitments that are necessary. Marine Corps representatives in particular noted that the extensive number of meetings occurring as part of the joint concept development and experimentation process was taxing its limited staff resources.

Thus, within the limited resources available, there is a need to obtain the proper balance between joint and Service-specific experimentation and to establish the appropriate priorities among all the various activities. At present, no formal mechanism exists for accomplishing this objective. The matter of achieving balance, by no means an easy task, is mentioned briefly at the end of this chapter and is addressed in detail in Chapter 6.

Historical Perspective

The Navy participated in numerous joint experiments prior to MC 02. MC 00, JEFX-1999, and JEFX-1998 are a few examples at the operational level. Outside the specific USJFCOM context, the Navy air arm has long participated in joint training exercises with USAF, including some experimentation that was transparent and nondetrimental to the exercise training objectives; the USAF tactical Red and Green Flag exercises at Nellis Air Force Base are examples. In addition, the Navy has a long tradition of wargaming at the operational and strategic level at Newport, Rhode Island; these war games have featured joint participation. The Navy has also participated in other Services' operational and war games ranging from the operational to the strategic.

TABLE 4.3 Navy and Marine Corps Initiatives in Millennium Challenge '02

Type of Initiative	Specific Initiatives in MC 02
Technical initiatives	Joint Fires (Navy Warfare Development Command (NWDC) sponsor)
Service-level integration initiatives	High-Speed Vessel (HSV) (NWDC and MCCDC sponsors plus others)
Navy Service initiatives	Joint and Maritime Component Commander Ship-Based Joint Command and Control Netted Force (distributed C2, coalition) Naval Fires Network—experimental Unmanned Sensors, Unmanned Platforms Theater Air Missile Defense Antisubmarine Warfare Antisurface Warfare Mine Warfare Information Operations
Marine Corps Service initiatives	Urban Combined Arms Exercise Local Area Security System Universal Combined Arms Targeting System

SOURCE: Derived from U.S. Joint Forces Command. 2002. "Millennium Challenge 2002, Forging Our Nation's Future Joint Force," Norfolk, Va.

Millennium Challenge '02

By far the largest joint experiment to date, Millennium Challenge '02 deserves special attention. The Navy's Fleet Battle Experiment-Juliet (FBE-J) and the Marine Corps Millennium Dragon 2002 experiment were conducted within the general framework of MC 02. They were quite large experiments in their own right. In terms of naval forces, MC 02 involved a Navy battle group and a Marine expeditionary brigade, as well as the extensive simulated representation of other forces. MC 02 initiatives involving Navy and Marine Corps sponsorship are shown in Table 4.3.

Initial Assessments of Naval Participation in Millennium Challenge '02

Table 4.4 presents, in highly summarized form, the Navy's "QuickLook"[22] assessment of Navy-specific objectives in MC 02. These are the only results

[22]Commander, Navy Warfare Development Command. 2002. "Naval Message to Commander, Fleet Forces Command, Norfolk, Va., Regarding QuickLook Report Fleet Battle Experiment-Juliet," NWDC, Newport, R.I., August.

TABLE 4.4 Navy "QuickLook" Assessment of Navy-Specific Objectives in MC 02

Navy Initiative	Comments in Assessment
Joint and Maritime Component Commander (JMCC)	Further development of JMCC doctrine required.
Ship-Based Joint Command and Control	Bandwidth limits explored.
Netted Force (distributed C2, coalition)	TBD
Naval Fires Network—experimental	Tactical Exploitation System-Navy (TES-N) and USAF ISR-M (the Air Force intelligence, surveillance, and reconnaissance data management system) interoperability demonstrated.
Unmanned Sensors and Platforms	Demonstrated
Theater Air Missile Defense	Requires common upper systems models and specific training skills.
Antisubmarine Warfare	Cable and hydrophone problems.
Antisurface Warfare	TBD
Mine Warfare	Long mission duration; Naval Mine Warfare Simulation requires subject-matter expert support.
Information Operations	Synchronization between components; no tools for e-kill battle damage assessment.

NOTE: Information on initiatives labeled "TBD" (to be determined) was unavailable at the time this report was written. SOURCE: Prepared by the committee on the basis of its reading of the Navy's "QuickLook Report for Fleet Battle Experiment-Juliet," see Commander, Navy Warfare Development Command, 2002, "Naval Message to Commander, Fleet Forces Command, Norfolk, Va., Regarding QuickLook Report for Fleet Battle Experiment-Juliet," NWDC, Newport, R.I.

known to the committee as of this writing. Since the analysis of MC 02 results has just begun, it is too early to discuss or comment on this assessment further.

A few major points were highlighted in the "QuickLook Report" for FBE-J. The most relevant of these are captured in the following paragraphs.

First, the high-speed vessel (HSV) was commended for its flexibility, speed, and modular design. The report states that additional experimentation and operational integration are "clearly warranted" and recommends that a second such vessel be leased and its capabilities assessed across the full range of littoral operations.

Second, the "QuickLook Report" strongly recommends a sweeping upgrade of the existing modeling and simulation environment, stating that "both the Navy and the Joint Community must improve the distributed simulation environment," since simulation is a vital augmentation even of these very large exercises. Unfortunately, the current modeling and simulation federation cannot simulate the emerging network-centric environment, creating serious limitations in the exploration of advanced concepts in MC 02. The "QuickLook Report" recommends either that the entire joint community move to a single (new) simulation environment, or that the existing environment be "radically improved."

Finally, the report notes that meaningful deployment of network-centric warfare concepts has been held back by lack of sufficient bandwidth available to deployed forces, but it notes that NWDC, with the help of the Naval Research Laboratory, has provided experimental links (via commercial satellite service) of 2 to 4 megabits per second to surface combatants and as much as 54 Mbps to a command ship. The committee continues to believe that per-ship bandwidth is a pressing concern, and it is extremely supportive of these experimentation efforts. NWDC deserves great credit for its ongoing and highly skillful work in this arena.

Committee Observations on Naval Participation in Millennium Challenge

On the basis of briefings and other information provided from the Navy, the Marine Corps, and USJFCOM, the committee makes the following, general observations with respect to Millennium Challenge '02:

- The value of such a large experiment relative to its cost was questioned both by a number of briefers and by many committee members. Some felt that such a large sum could be put to better use in a series of smaller experiments both of the joint and the Service-specific type. Others noted, however, that such an event made the assets of other Services available, which is difficult to accomplish in the context of smaller experiments because of all the demands on operational assets. There was full agreement that it is extremely important to select the venue that matches the objectives desired. A large-scale field experiment may be warranted to test integration, scalability, and a complex set of interactions. But the question remains—How large a scale is necessary?
- Such large, high-visibility experiments were characterized by many as demonstrations, rather than experiments. In demonstrations there is little room for true exploration or failure. If that is the case, such events may provide some opportunities to showcase unfunded but "ready-for-prime-time" equipment capabilities in hopes of garnering support and supplemental funding. Others noted, however, that the series of smaller events conducted over the 2 years leading up to MC 02 allowed for greater exploration and assessment.
- The greater use of simulation relative to the use of live forces would allow more experimentation in the future, especially for concepts that apply above the tactical level. For example, the Standing Joint Force Headquarters, one of the main concepts tested in MC 02, could perhaps have been adequately explored with few if any live forces or real platforms.

• Ongoing naval participation is essential during the planning stages for major joint experiments in order to ensure that Service experiment objectives are included in the formal list of joint experiment objectives. In this respect, the Navy and the Marine Corps have done well to date; in MC 02, the Navy had 10 Service objectives and the Marine Corps had 3 included in the joint experiment plan.

Joint Experimentation at the Tactical Level

A final perspective for looking at naval involvement in joint experimentation is that of the command level involved: that is, the Joint Task Force (JTF) level, component (maritime, land, or air) level, or tactical level. In joint experiments to date, most joint participation of the Services has been at the JTF level, with some lesser degree at the component level. This raises an important question for the future: Can and should greater concept development and experimentation be conducted jointly at the tactical level? It appears that this will be necessary, given the joint interaction that was witnessed in tactical operations in Operation Enduring Freedom in Afghanistan (e.g., Army Special Operations Forces directing fire from Navy and Air Force aircraft).

It is yet to be determined how best to accomplish greater concept development and experimentation conducted jointly at the tactical level. The committee notes that USJFCOM appears to be pushing downward by influencing experimentation at levels below its original JTF perspective, although perhaps mostly only to the component level. The Services appear concerned about losing their prerogatives in this way, and with some reason. Perhaps this is a matter to be worked out in cross-Service (vice joint) forums.

The Price of Naval Participation in Joint Experiments

Experimentation is an important part of transformation—in fact it might be viewed as the engine of change—but experimentation certainly comes with a price tag. Many elements of this "cost" can readily be quantified in dollar terms, but the cost of other very real elements can be difficult to quantify. One important area is the cost in terms of human resources.

The number of personnel needed for the actual execution of an experiment is easily visible, but the human resource requirement for experiment *planning* is less so. Not only are there a variety of technical activities associated with the design of the experiment, but there are also numerous coordination activities. For example, Figure 4.3 indicates that more than 100 separate events are scheduled for joint experimentation in the FY 2004–FY 2009 period. Each of these will most likely require numerous meetings and related activities. The limits imposed by resource constraints and the demands of real-world operations tempo mean that the personnel demands cannot be easily met. However, the situation does argue for the joint and Service communities to confront this matter as judiciously as possible—by avoiding unnecessary or redundant activities and holding experiments to the minimum size necessary to accomplish their objectives.

EXPERIMENTATION IN THE COMBATANT COMMANDS

The November 2002 guidance issued by the Chairman of the Joint Chiefs of Staff, as previously quoted, states that "[USJFCOM's Joint Experimentation Campaign Plan] must incorporate a decentralized process to explore and advance emerging joint concepts, proposed operational architectures, experimentation and exercise activities currently being conducted by . . . the combatant commands [and specified other entities]."[23] Thus, while USJFCOM is the executive agent for joint experimentation, such experimentation also takes place in the Combatant Commands beyond those experiments directly organized by USJFCOM. In fact, the issues involved in joint force operations are so important and large in scope that it is only natural that all Combatant Commands should be involved in addressing them through experimentation.

The primary responsibilities of a Combatant Command can be summarized as follows:

- *Readiness*—being prepared to respond to contingencies, from small crises to major conflict;
- *Security cooperation*—conducting activities to build regional coalition capabilities to carry out common missions; and
- *Transformation*—adapting force operations to take best advantage of new concepts and technology to meet existing threats and to be prepared for possible future threats.[24]

While the first two items above have long been considered the responsibilities of a Combatant Command, today's world environment has also elevated the third to key importance. The Combatant Commands must be prepared to meet the threats evolving today in the theaters. In this regard, the Combatant Commands bring a unique perspective complementary to other transformation efforts in the DOD—that is, their understanding of emerging and potential threats in terms of the concrete aspects of a specific geopolitical context.

---

[23]Gen Richard B. Myers, USAF, Chairman, Joint Chiefs of Staff. 2002. "Guidance for USJFCOM Joint Experimentation [Enclosure]," Memorandum for the Commander, U.S. Joint Forces Command, November 26, pp. 1-2.

[24]Derived from comments made by ADM Dennis C. Blair, USN. See "Statement of Admiral Dennis C. Blair, U.S. Navy, Commander in Chief, U.S. Pacific Command, before the Senate Armed Services Committee on U.S. Pacific Command Posture," March 27, 2001.

Combatant Command exercises (which may involve war games and simulations as well as live forces) provide a major vehicle for addressing all three responsibilities: Exercises train U.S. forces, thereby helping to maintain readiness. When involving coalition forces, they help build cooperative operational procedures among U.S. and coalition partners. And when involving an element of experimentation, these exercises also lead toward transformation.

The inclusion of experimentation within exercises warrants further comment. Some look upon this negatively while others favor it. The following quotation gives one Combatant Commander's perspective:

> Some have contended there must be a bright and shining line between training and experimentation. They say too much experimentation during an exercise on the one hand degrades the training and on the other hand constrains the development of truly revolutionary leaps forward in warfighting concepts. I believe that experimentation meshes well with training. We should incorporate an element of experimentation in our exercises in a way that will accomplish the goals of both and often will enhance both. This is especially true for sensors and communications systems, the keys to information dominance.[25]

Given that Combatant Commands today must address all three responsibilities listed above—readiness, coalition collaboration, and transformation—the committee finds itself in general agreement with this view. The need to meet all three responsibilities and the limited time and assets available with which to do so means that, realistically, some experimentation must take place within a broader context of exercises. While an experiment conducted within the context of an exercise may not be able to have all the discipline of a "pure experiment," it should, if conducted correctly, allow for new concepts and technologies to be explored, with the lessons from that exploration being captured and acted upon. The exercise context will require that the transformational capabilities being addressed are largely evolutionary, but certain aspects of transformation may well be built from a sequence of evolutionary steps rather than from one bold revolutionary leap.

When combining experiments with exercises, it is crucial to preserve the objectives of both, although this has often been difficult.[26] The committee believes

---

[25]ADM Dennis C. Blair, USN. 2001. "Change Is Possible and Imperative," *U.S. Naval Institute Proceedings,* Vol. 127, No. 5, May, p. 49. ADM Blair also goes farther with this point in a way that is relevant to this report as a whole. Namely, he argues for "acquisition by adaptation," whereby a prototype system is put out quickly and adapted and improved as it is fielded, through such venues as exercises.

[26]See, for instance David S. Alberts, Richard E. Hayes, John E. Kirzl, Leedom K. Dennis, and Daniel T. Maxwell, 2002, *Code of Best Practice Experimentation,* DOD Command and Control Research Program, Office of the Assistant Secretary of Defense (Networks and Information Integration), Washington, D.C., July, Ch. 3, p. 345. Available online at <http://www.dodccrp.org/>. Accessed October 7, 2003.

that success requires that the experimental objectives, design, conduct, and data collection and analysis must be as rigorous as possible, and that they not be compromised. Success also requires that planning take a "campaign" perspective, since many of the experimental objectives will entail a series of exercises in order to be realized.

To make the points discussed here more concrete, Table 4.5 presents several examples of experiments conducted by Combatant Commands. Some general features of these examples are apparent. First, the experimentation applies largely to command, control, and communications. Second, it takes place in both command post and field exercises, and it pertains both to developing and refining procedures and to testing and refining prototype systems. Third, the experimentation can involve coalition nations, which is significant because future operations—from humanitarian actions through major conflict—would most likely involve coalition partners.

There are numerous examples, such as those cited in Table 4.5. However, there does not appear to be any coordinated program between the Services and Combatant Commands to conduct experimentation within joint exercises in a systematic way.[27] Such coordination would involve the organized execution of the following activities by the Service and Combatant Command staffs:

- Determining cases in which participation in such field events is the most effective venue;
- Identifying areas of joint operational concepts and tactics (e.g., based on the essential tasks of a joint mission) requiring further development in order to best meet challenges anticipated in the theater;[28]
- Identifying Service systems under development that are suitable for examination in experiments;
- Planning for exercises to most fully accommodate experimentation involving these concepts and tactics and systems, consistent with the other objectives of the exercises; this planning should assume a campaign perspective, since many of the experimental objectives will require the development of a body of knowledge through a series of activities;
- Conducting the experimental design, data collection, and data analysis necessary to make the experimentation as rigorous as possible;

[27] ADM Dennis C. Blair states, "Currently efforts like this are ad hoc. What is needed is a coordinated system between the services and Combatant Commanders in which we incorporate emerging systems into major CPXs and FTXs to ensure they are meeting real current and future needs." ADM Dennis C. Blair, USN. 2001. "Change Is Possible and Imperative," *U.S. Naval Institute Proceedings*, Vol. 127, No. 5, May, p. 48.

[28] Likely areas for examination include establishing a common operating air picture and coordinating joint firepower or—to pick an emerging threat of particular concern—countering swarm tactics in littoral areas.

TABLE 4.5 Examples of Experimentation in Combatant Commands

Event	Nature of Experimentation
Command and Control Exercise (Pacific Command (PACOM), twice yearly)	Examined information flows between Tier II and Tier III of Joint Task Force (JTF) Headquarters, e.g., involving common operating picture, air tasking orders, and requests for fire power. Improved ability of Tier III components to operate as part of the JTF.
Kernel Blitz (Experimental) (PACOM, 2001)	Run in conjunction with Fleet Battle Experiment-India (FBE-I) and Capable Warrior AWE; involved forces from all four Services. Explored Wide-Area Relay Network (WARNET) technologies. Supported development of JTF WARNET to provide a secure tactical communications network.
Foal Eagle (PACOM, 1998)	Run in conjunction with FBE-Delta, incorporating Army assets into that Navy experiment. Demonstrated concept of using Army Apache helicopters against Special Operations Forces infiltration craft, and left behind a system for linking Navy and Army fire control systems.
Internal Look (Central Command (CENTCOM), 2002)	Tested a deployable joint command and control facility. Helped shake out new capability by examining procedures and communications from theater to continental United States and within theater.
RIMPAC (PACOM, 2000)	Combined exercise involving seven countries (that rim the Pacific Ocean). Established satellite communications for a coalition wide-area network involving all 50 participating ships, providing connectivity to command ship and data sharing among all ships. Enabled enhanced combined operations.
Cobra Gold (PACOM, yearly)	Combined exercise with Thailand; also some involvement from Singapore. Helped to work out coalition procedures and networks.
Advanced Concept Technology Demonstrations (ACTDs)	Each ACTD[a] has a Combatant Command as its operational sponsor, and exercises can be used to explore and refine the capabilities involved. For example, the Extending the Littoral Battlespace and Commander in Chief (CINC 21) ACTDs, which are providing new C3 capabilities, participated in Kernel Blitz (Experimental).

[a]There have been more than 100 ACTDs since the inception of such programs in 1995.

- Disseminating relevant results of the experimentation (perhaps through USJFCOM) to other Combatant Commands, and considering how in a broader sense the results might influence joint doctrine; and
- Feeding the results of the experimentation on developmental systems back into the development process (i.e., to support spiral development).

The most direct mechanism for this coordination would be between the Combatant Command and the Service component commands assigned to that Combatant Command (e.g., either the appropriate Navy fleet command or the numbered fleet commands under that fleet command). The component commands could then interact as necessary with other elements of their respective Service (e.g., with NWDC and MCCDC).

In summary, the Navy and Marine Corps and the Combatant Commands cooperate in joint experimentation within exercises conducted in the Combatant Commands—Kernel Blitz is one such example. However, it appears highly desirable that there be more active collaboration of the Navy and Marine Corps and the Combatant Commands to systematically develop programs of joint experimentation. Both large and smaller exercises conducted in the course of routine deployments could be involved, as would both fleet and command post exercises. While there are limitations to conducting experiments within exercises because of the multiple objectives that the exercises themselves must satisfy, there are also numerous benefits to be realized. In particular, the exercises can offer frequent opportunities for joint interaction, they can lead to results being fed directly back into the operational forces, they can support the spiral development process through joint (rather than Service-unique) system use, and they can allow the development of concepts with coalition partners.

Fleet battle experiments have been involved to a limited extent in joint experimentation—for example, as noted in Table 4.5 regarding FBE-D and FBE-I. It appears that the much greater involvement of FBEs in joint experimentation would be a particular opportunity for the Navy to enhance its joint capabilities and to contribute to joint capabilities overall. The FBEs are run under the numbered fleets and, as noted above, those commands provide the most direct vehicle for interacting with the regional Combatant Commands.

CROSS-SERVICE EXPERIMENTATION

Not all experimentation involving two or more Services need take place under the joint umbrella. Although joint leadership of multi-Service experiments should ensure that the experimentation relates to joint concepts and supports joint doctrine development, direct Service interaction can involve fewer complications and costs and greater freedom to innovate. These are obviously advantages. Any useful concepts developed in such direct Service interaction should, however, eventually be fed into the joint arena.

In the committee's view, concept development and experimentation involving the Services directly is best suited for operations at the tactical level, since higher-level (component and JTF) operations are primarily joint. This direct Service interaction can occur in two general ways—between Service centers involved in concept development and between deployed forces engaged in exercises and experiments. Representatives from Service centers are brought together during joint activities (e.g., Millennium Challenge), but the reason is to work toward some generally predefined purpose; what is envisioned here are interactions allowing greater freedom for the exchange of ideas. In that regard, the Navy Warfare Development Command and the Marine Corps Combat Development Command have regular interaction with one another. In addition, the Navy Network Warfare Command has noted that its proximity to corresponding Air Force organizations (e.g., Air Combat Command and the Air Force Command, Control, Intelligence, Surveillance, and Reconnaissance Center) offers the opportunity for collaboration. While other, substantive interactions between Service centers may exist, none were apparent to the committee in its investigations. Increased interaction could be very valuable—to provide cross-fertilization in concept development and to begin building joint concepts from the ground up.

Regarding direct Service interaction in exercises and experiments, the Navy indicated in briefings to the committee that other Services (e.g., the Air Force) are involved in its fleet exercises, and that it would like to increase such involvement, although mutual scheduling is a difficulty. Recent Marine Corps experiments, with their highly tactical focus, do not appear to have a significant joint or cross-Service perspective. However, future opportunities would appear to exist—for example, coordinated operations with Army Special Operations Forces and air support from Navy and Air Force aircraft.

As just noted, one recurrent difficulty with involving another Service in Navy fleet exercises is that of arranging the simultaneous availability of assets from the two Services. Ideally, the elements of both Services should be on exercises (or other, more formal experiments) at the same time. This, however, requires cross-Service coordination of force deployment schedules. Such coordination would have to be carried out at the top levels of the Services—possibly with USJFCOM acting as an intermediary—because the deployment schedules have Service-wide impacts relating to such factors as accomplishing operational missions, training, personnel rotation, and platform overhaul and upgrade.

ENHANCING NAVAL PARTICIPATION IN JOINT EXPERIMENTS

Now is a time of great ferment in military thinking, both within the Navy and the Marine Corps and across the joint community. Changes in the systems for defense requirements generation and acquisition are actively being promoted, and a transformation process is being put into place to develop new military capabilities more rapidly in order to be prepared for future military challenges. The lessons from Operation Iraqi Freedom will further shape the understanding of needed capabilities. Concept development and experimentation figure as a key element in all of these initiatives. Furthermore, since all recent military operations have been joint, and most future operations will likely be so as well, the joint aspects of concept development and experimentation are particularly important.

As this chapter describes, the Navy and the Marine Corps have been active participants in joint concept development and experimentation. However, both the need and the opportunities for greater participation are also recognized here. While the committee's perspective is primarily that of the Navy and Marine Corps, it is recognized that such participation is a two-way collaboration—that is, the Navy and Marine Corps work toward joint objectives, and the joint authorities are likewise receptive to the particular ideas that the Navy and Marine Corps bring.

The need for naval-specific concept development and experimentation is a key factor affecting naval participation in the joint counterpart, given the limited resources available and the recent strain caused by large-scale joint experimentation. Thus, the proper balance between naval-specific and joint activities must be determined. A set of principles and guidelines is necessary to help achieve this balance. These are provided as part of the committee's recommendations in Chapter 6, in concert with a mission-based strategy—a process to determine where to focus joint and naval experimentation, respectively.

5

Effectiveness of Experimentation for Future Naval Capabilities

This chapter assesses the effectiveness of experimentation for future naval capabilities. It addresses the specific questions of the terms of reference, beginning with a review of what has been learned from naval experimentation and an assessment of the success of transitioning results from experiments to the fleet and field. The assessment then naturally leads to an examination of the adequacy of naval experimentation programs, processes used, and the environment and infrastructure that support experimentation. The chapter concludes with an examination of the broader relationship of experimentation within the joint context, specifically looking at the effectiveness of naval experimentation in preparing for joint operations and at the relationship between Service-unique and joint experimentation.

ASSESSMENT OF EXPERIMENTATION RESULTS

The terms of reference ask what has been learned from experimentation thus far. From a historical perspective, experimentation with innovative technologies in military systems and with techniques of warfare has often led to revolutionary changes in how military forces are constituted and how they fight. Examples from the previous century include the use of armored forces in land warfare, the use of submarines and sea-based aviation in naval warfare, the application of ballistic missiles to intercontinental attack capability, the use of nuclear submarines to constitute long-enduring and essentially invulnerable undersea tactical and strategic strike forces, and the use of space systems for observation, communication, and navigation.

The Navy and Marine Corps have a long history of using experimentation to evolve significant new capabilities. The Navy experimented with submarines, carriers, and PT boats, and with new propulsion systems and new fuels. Experimentation with launching and recovering aircraft from ships began but a few years after aircraft were invented. Similarly, although Marines had been landing from ships in small boats for many decades, even centuries before World War II, the pressures of warfare led to the rapid development of prototypical modern amphibious landing systems. Experimentation has been key to advances in all forms of naval warfare—gunnery, guided missiles, naval ship propulsion, vertical-capable jet aircraft, very short takeoff and landing rotorcraft, and all other activities related to the shaping and operation of naval forces, including the Fleet Antiterrorism Security Teams and the Chemical/Biological Incident Response Force (CBIRF) of today.

In order to assess more recent Service-unique experimentation, the committee focused primarily on the Navy Warfare Development Command (NWDC)-sponsored fleet battle experiments[1] (Alpha through India) and on the Marine Corps experimentation efforts beginning with the Hunter Warrior (HW), Urban Warrior (UW), and Capable Warrior (CW) campaigns. Fleet Battle Experiment-Alpha (FBE-A) was conducted in March 1997. The Hunter Warrior series of activities was initiated in 1997. Chapter 3 provides details on these efforts.

FBE-A through FBE-I included a total of nearly 40 separate objectives. Each FBE had between three and eight major objectives, and each major objective had anywhere from one to nine subobjectives. A significant number of these major objectives and subobjectives were realized. The range of investigation of these FBEs was quite extensive, addressing network-centric operations for naval and joint fire power, theater and air missile defense, precision engagement, time critical strike, and defense against asymmetric threats, to name a few areas. Table 3.1 in Chapter 3 provides a synopsis of all FBEs, their objectives, and their results. Collectively assessed, these provide evidence that experimentation is achieving meaningful results.

The three experimentation campaigns HW, UW, and CW also addressed many objectives. HW had 37 objectives, of which 29 were realized. As a campaign, it focused on individuals and combat patrols operating in desert environments. UW addressed individuals and platoon-size operations in urban environments; CW focused on individuals and company-size operations at Camp Pendleton, California. Each of the campaigns required a cycle of more than 3 years. As with the FBEs, these campaigns covered a considerable range in their investigations,

[1]The nominations for and participation in FBEs involve many organizations, as discussed in Chapter 3. For example, these organizations have included the Second, Third, Fourth, Fifth, Sixth, and Seventh Fleets, and regional commanders, who propose experiments through the Navy Component Commanders in their area of responsibility.

including the use of nonlethal weapons, small unmanned aerial vehicles (UAVs), sensors, intrasquad radios, precision targeting, and the development of a common tactical picture. Each addressed a series of important questions, such as how to extend the effectiveness of a modest forward afloat expeditionary force or how to penetrate and operate in dense urban littorals. In short, many objectives of each campaign were realized and valuable knowledge was derived. The lessons learned from these campaigns were captured in the Marine Corps "X-files."[2]

To the terms-of-reference question posed at the beginning of this section, the committee responds that both Navy and Marine Corps experimentation is enabling learning and producing meaningful results directed at promising concepts and technologies in a number of key naval mission areas. However, other questions that remain to be answered in this chapter go to the heart of the matter—which is whether naval experimentation is as effective as it can be and needs to be.

ASSESSMENT OF TRANSITIONING

A key question from the terms of reference is how successful the transitioning of the results of experimentation to the field has been. The committee responds by focusing on (1) doctrine and tactics, techniques, and procedures (TTPs) and (2) fielded capabilities, including acquisition programs. Summaries are provided in Table 5.1, which synopsizes results from the discussions on naval experimentation in Chapter 3. More detailed observations are presented below.

Summary Observations

One of the important outcomes of the Navy's FBE series of experiments has been the determination as to why certain objectives were not achieved, which allows for an iterative process of improvement. The principal successes of FBEs have been as follows:

---

[2]"The X-Files contain useful information packaged for rapid reading and easy transport in the cargo pocket of the utility uniform. They convey a synthesis of knowledge gained from experiments with tactics, techniques, and procedures, and some enabling technologies that can help us fight and win battles. Most of them focus on operation in the urban battlespace. They are an evolving body of knowledge that is constantly refined through experimentation. . . . information in the X-Files is entered into the Marine Corps Combat Development System. It forms the backbone of recommended revisions to Marine Corps doctrine for Military Operations on Urbanized Terrain (MOUT). Knowledge in the X-Files also underpins much of the Basic Urban Skills Training (BUST) program used by the Operating Forces. The X-Files gather, organize, and synthesize knowledge from post training analysis and feedback from Marines, Sailors and other participants in MCWL experiments. They do not contain official doctrine, nor are they policy or standing operating procedures (SOPs)." For further information see <http://www.mcwl.quantico.usmc.mil/x_files.asp>. Accessed November 18, 2003.

TABLE 5.1 Transitioning Results and Conclusions of Experimentation into the Navy and Marine Corps

Result	U.S. Navy	U.S. Marine Corps
Doctrine and tactics, techniques, and procedures (TTPs)	Some concepts have moved into the field	Many concepts have moved into the field.
Fielded capabilities including acquisition programs	Few successes so far. Naval Fires Network is one notable concept that appears to be in transition to the field.	Numerous successes in transitioning small end items as well as interim capabilities.
	Transitions are very difficult, and processes for achieving them are seen as poor, even from within the Navy, owing in part to budget pressures and to a lack of processes for new capabilities to compete with programs of record.	Transitions have proven very difficult for expensive capabilities because of budget pressures and ineffective processes for new capabilities to compete with the programs of record.

- Demonstrations of the feasibility of new operational concepts using surrogate or prototype or existing systems,
- The adoption by fleet forces of new TTPs, and
- The development of new doctrine for fleet operations.[3]

The Naval Fires Network (NFN) appears to be a case of successful experimentation in the process of transition for the Navy. NFN is a network-centric warfare system that provides real-time intelligence correlation, sensor control, target generation, mission planning, and battle damage assessment capabilities. It allows ships and aircraft in a carrier strike group or an expeditionary strike group to share near-real-time and real-time intelligence and targeting information, not only with one another but also with Army and Air Force units in a joint or coalition task force. NFN was first studied in FBE-A and then refined in FBE-I. It has transitioned to the fleet as an interim prototype on two carrier strike groups, owing to intense interest from the CNO and fleet commanders. However, to date it has not become a formal program of record.

Marine Corps successes, based on the Hunter Warrior, Urban Warrior, and Capable Warrior campaigns, are as follows:

---

[3]In Chapter 3, see the section entitled "Synopsis of Results to Date from Fleet Battle Experiments Alpha Through India."

- Concepts, doctrine, and TTPs resulting from experimentation have transitioned successfully to forces in the field.
- Experimentation has resulted in changes in minor equipment items in the field.[4]

It is clear that naval experimentation is resulting in new doctrine, TTPs, some new concepts, and some minor end items. It is also clear that there is a serious shortfall in transitioning the results of experimentation into major fielded capabilities. For the Navy, this extends to minor items of equipment and capabilities as well. The committee's findings with respect to the success of transitioning results of experimentation to the field are as follows:

Finding for Navy: The mechanisms and processes for transitioning the results of experimentation directly to the fleet or to an acquisition program of record are inadequate, and they curtail the effectiveness of experimentation in building future naval forces.

Finding for Marine Corps: The Marine Corps has been successful in transitioning nonmaterial elements of doctrine, organization, training, materiel, leadership, personnel, and facilities (DOTMLPF) and minor equipment. However, it has not been successful in transitioning to combat forces major warfighting capabilities identified during experimentation.

More detailed discussions related to these findings follow.

Transitioning to Naval Doctrine and Tactics, Techniques, and Procedures

One objective for experimentation is to explore or assess new doctrinal concepts and, later, to develop and refine new TTPs. The committee noted that the NWDC is sensitive to the need to coevolve doctrine and TTPs at the same time that new technology is introduced. The Marine Corps Combat Development Command (MCCDC) also emphasizes this strategy—partly because in some early experiments problems had arisen as a result of insufficient attention to retraining participants in the use of new capabilities. The lessons of the Gulf War in 1991 and of the conflicts in Bosnia and Kosovo (where sustained operations required superb command, control, communications, computers, intelligence, surveillance, and reconnaissance (C4ISR)), and in Afghanistan (where Special Forces and Marines operated directly together with the use of long-range bombers) were consistent with studies and games conducted over the years. Without the urgency of an actual war, however, it is unclear that change would have happened,

[4]In Chapter 3, see the subsection entitled "Synopsis of Results of Sea Dragon to Date."

since at a certain stage, rational change driven by persuasive analysis, modeling, simulation, and gaming is not yet supported by empirical evidence. Nonetheless, the "wild ideas" practiced in experiments such as Hunter Warrior, Urban Warrior, and fleet battle experiments were turned into reality when the time came—because individuals exposed to the ideas and potentials were willing to break with tradition and doctrine to do sensible things. That said, some results produced from Navy and Marine Corps experimentation have led to new concepts, doctrine, and TTPs, as well as new training initiatives.

FBEs have been successful in the process of developing and transitioning into the fleet decision-support concepts and tools intended to increase the speed of command. Other FBEs enabled the collaboration of command echelon decision makers. One specific effort is the invention of a Knowledge Web (K-Web) concept.

As described in Chapter 3, the K-Web involves the application of knowledge-management practices to warfighting, creating a concept of operations in which value-added information (i.e., knowledge) is created and published on the command intranet in real time rather than being coupled to daily briefing cycles. The concept was initially developed in war games, installed as a prototype system on the USS *Carl Vinson*, and eventually battle-tested during Operation Enduring Freedom, where it was viewed very favorably. The Network-Centric Information Center is working to migrate these tools and transition them to other battle groups. The K-Web demonstrates a significant first step toward a new concept of operation for warfighting and is explicitly designed to support distributed collaboration.

Other examples of successful transitions resulted from the Marine Corps Hunter Warrior, Urban Warrior, and Capable Warrior campaigns, which shaped doctrine and TTPs in the field in the following areas:

- Introduction of new command relationships,
- Development of the Combat Decision Range,
- Construction and use of the urban Close Air Support range (Yodaville),
- Development of military operations in urban terrain (MOUT) for infantry tactics,
- Continued refinement of the CBIRF organizational structure, and
- Construction of the night laboratory for training infantry tactics.

In particular, the results of Urban Warrior have influenced major changes in the doctrine of how to fight in urban terrain. Additionally, many of the training lessons learned have been captured in the X-files, which are distributed throughout the Marine Corps down to the squad level. The development and building of the night laboratory for the training of rifle squads and all lieutenants at the Basic School (the primary school of basic training for all Services) are a direct result of HW and UW. Another training tool, the Combat Decision Range, is also a direct product of the experimentation program. The use of paintball and chalk small arms rounds for urban force-on-force training is also a result of the Marine Corps experiments in urban warfare. The instrumented Close Air Support range (Yodaville) constructed near Yuma, Arizona, is a direct result of UW and is currently used by Air Force, Navy, and Marine Corps pilots.

Transitioning Results to Acquisitions and Fielded Capabilities

In reviewing the experimentation efforts of the NWDC and the MCCDC, the committee concludes that there have not been any *major* end items of equipment that have moved directly into acquisition as a result of experimentation.

Navy

The Navy's process for turning concept development over to the acquisition system is viewed from within as being poor[5] (discussed in more detail later in this chapter). Instead, one sees examples of the Navy moving directly to the procurement of interim capabilities, sometimes prematurely. Some critics of the Naval Fires Network believe that it was rushed into the fleet with insufficient consideration of alternative ways to achieve the same ends, while supporters view the movement as an opportunity to understand the NFN's potential and to use that knowledge to guide its further development and effectiveness.

However, as noted in Chapter 3, Navy experimentation results (e.g., with the high-speed vessel[6]) have influenced some acquisitions, such as the Littoral Combat Ship. The route of entry seems to be through interested individuals in leadership positions, evolutionary upgrades of existing systems, or modification of the design concept or implementation plans of ongoing programs of record.

The NFN has the potential for the most successful transition of the results of an FBE-tested concept, although it is not yet a formal program of record. The successful test of the NFN in FBE-I and its subsequent introduction into the fleet as an interim prototype on two carrier strike groups demonstrate that experimentation may be used as a vehicle for the rapid introduction of new capabilities into the fleet.

---

[5]Admiral Dennis Blair expressed his concerns on this matter while still serving as U.S. Commander in Chief, Pacific (see ADM Dennis C. Blair, USN, 2002, "Force Transformation in the Pacific," remarks at U.S. Naval Institute/Armed Forces Communications and Electronics Association Western Conference 2002, San Diego, Calif., January 15, pointing to examples of systems readily available in the civilian economy that have not yet been adapted for use in the operating fleets.

[6]As with its simulated use in FBE-J, the high-speed vessel is serving as a command and control platform and staging base in Operation Iraqi Freedom. See Jason Ma, 2003, "Catamaran Deployed in War: Naval Special OPS Use Joint Venture in Operation Iraqi Freedom," *Inside the Navy*, Vol. 16, No. 13, March 31, p. 1.

Marine Corps

The committee observed that the Marine Corps has had some successes in getting interim minor end-item equipment capabilities fielded as a result of experimentation—for instance:

- Nonlethal weapons sets for deploying Marine Expeditionary Units,
- Laser eye protection kits for deploying units,
- PAQ-4 night alignment sight for the M-16,
- Intrasquad handheld radios for all USMC units,
- Elbow and knee pads for MOUT operations, and
- Infrared-treated battle dress uniforms.

Successful results of experimentation for major equipment items have been very difficult, if not impossible, to transition to fielded capabilities. Examples of successfully transitioned results include:

- Interim fast attack vehicles,
- Precision targeting systems for forward observer/forward air controllers (FOFACs), and
- The Dragon Eye UAV.

The interim fast attack vehicle (discussed in Chapter 3) has been fielded to the operating forces, while the program of record—the light strike vehicle—works its way through the acquisition system. To the committee's knowledge, the debate on displacing the program of record did not occur. The commercial off-the-shelf (COTS) intrasquad radio (experimented with during Urban Warrior) has been fielded as an interim capability, while the formal acquisition system works on the Joint Tactical Radio System.

The forward observer forward air controller system equipment, barely workable during Hunter Warrior, was improved in terms of its reliability, ruggedness, and performance during Urban Warrior; the prototype systems are now deployed with units. Meanwhile, the formal FOFAC program of record—that is, the Target Location and Designation Handoff System (TLDHS) and the Target Handoff System (THS)—is working its way through the aviation acquisition system.

Additional experimental equipment currently in operational use includes the Dragon Eye UAV, which was sent to the First Marine Expeditionary Force in Kuwait to support Operation Iraqi Freedom. This small reconnaissance and surveillance system provides real-time video imagery and can be put in a backpack. An acquisition program for the Dragon Eye is scheduled to follow its extended evaluation by users, although it has already undergone more limited development and operational evaluation, as well as participation in the Millennium Dragon '02 Marine Corps experiment.

Transition Planning

Logically, one might expect that well-designed and well-executed experimentation campaigns would be so persuasive that the transition of equipment from them into acquisition would be straightforward. In practice, this is clearly not the case. The Marine Corps instituted formal planning to facilitate transition. The Navy does not appear to have done so, relying on personal interactions by leaders to move capabilities forward.

The lack of a formal transition plan is only one impediment to successful transition. Success depends fundamentally on satisfying a requirements process and on having resources made available in advance. That is, if an experiment campaign ends gloriously but no funds have been provided in the Service Program Objectives Memorandum, exploiting the fruits of experimentation proves to be extremely difficult for obvious reasons. A solution should rely on planning and programming in advance and/or on invoking a process that amounts to a competition between the capabilities developed through experimentation and the program of record. Since the program of record represents the results of many compromises and a prioritization by the senior leadership, the notion of supplanting an agreed-to program with a new one emerging from an experiment is difficult. Since time is important, the transition process needs to start early—at the time the experiment campaign is planned, in anticipation of success.

Not all experimental capabilities warrant transitioning by displacing a program of record. Potential capabilities may remain promising but unrealized because their introduction into the field is not as effective a return on investment as is the capability already programmed. Yet if experimentation is to provide results to the fleet, the debate should be allowed. The present process is woefully inadequate; it almost guarantees that the results of successful experimentation will not enter the acquisition process or become the basis of a program of record.

ASSESSMENT OF SPIRAL DEVELOPMENT

There were two sources of motivation for considering spiral development in this report. First, the terms of reference for the study specifically asked the committee to do so. Second, the network-centric operations report of the Naval Studies Board recommended that the Navy and the Marine Corps apply spiral processes within the experimentation framework as a means to develop concepts and processes and to accelerate fielding capabilities.[7]

All of the groups delivering briefings and written materials to the committee referred to spiral development, but the term carried very different meanings.

---

[7]Naval Studies Board, National Research Council. 2000. *Network-Centric Naval Forces: A Transition Strategy for Enhancing Operational Capabilities,* National Academy Press, Washington, D.C.

Many individuals seemed knowledgeable about the spiral development process. In some cases, what was described as spiral development was actually traditional product improvement in phases.[8] Relatively little evidence was seen in naval briefings of an explicit planning process for true spiral development, which emphasizes iterative cycles of prototypes with substantial learning and adaptation based on operator feedback to drive subsequent spirals before delivering a fielded capability.

More typically experiments—such as those using commercially available high-speed vessels—amount to the use of a prototype or surrogate that approximates what will eventually be required. The information from one experiment featuring the vessel is used to shape a subsequent experiment and to answer additional questions. However, such a sequence of events does not constitute a deliberate spiral acquisition. The submarine community used a spiral methodology for the Advanced Rapid COTS Insertion (ARCI) program to address the (then) loss of acoustic advantage in U.S. submarines.[9] Another example of experimentation used to support a continuous evolutionary improvement is that of the FOFAC, cited above. Developed under Hunter Warrior, this system was the subject of enhancements during Urban Warrior and resulted in an interim fielded capability.

These things said, the committee was struck by the manner in which naval experimentation is disconnected from the Navy's acquisition community. In some briefings it was asserted that experimentation was not relevant to acquisition. Only in a few instances was there evidence of a deliberate, planned connection—to apply what was learned in experiments to shape system acquisitions or to turn the results into a coherent acquisition program.[10]

It appears as if some Navy interest in experimentation is really based on a desire to bypass the acquisition process. This is understandable, given the notoriously long times for normal system acquisition. But the price paid is high: By hopping from concept development to experiments and interim capabilities, the Navy is bypassing the organizations with the expertise to do systematic engineering, capabilities-based planning, and operational assessments. The committee also observed that the Navy's acquisition components seem to be viewed by some leaders as a major part of the problem, rather than as part of the solution. Nonetheless, the Naval Fires Network could be cited as an example of something that leaped quickly from concept demonstration to directed procurement. The

[8]Virtually all major systems have involved evolutionary development and deployment under names such as "block change" or "planned product improvement strategy."

[9]In Chapter 3, see the subsection entitled "A Recent Experimentation Program: Advanced Rapid Commercial Off-the-Shelf Insertion."

[10]The most prominent example cited is that of the Littoral Combat Ship, which is using experimentation with the high-speed vessel to determine key performance parameters.

basic idea is good and interim capabilities are much to be desired, but a refined design is probably needed for all but the near term. The NFN could be viewed as an example of a capability that could profit from a deliberate spiral development process managed collaboratively by the experimentation and acquisition communities. Such problems can be remedied in part by the advanced planning process described above, since the initial experimentation with the NFN would have been viewed as part of a campaign, with orderly provision for acquisition built in at the start.

The committee concluded that the spiral development process is used at the discretion of the individual manager and is confined to discrete experimentation events or individual acquisition capabilities rather than being applied systematically. An assessment of the effectiveness of this methodology could not be made because its use is too sporadic. The committee believes that the Navy and the Marine Corps have not yet adequately explored spiral processes or spiral development, particularly in the context of experimentation campaigns, as was recommended by the network-centric operations report of the Naval Studies Board.

ASSESSMENT OF THE NAVAL EXPERIMENTATION PROGRAM AND ITS METHODS

The terms of reference specifically ask the committee to determine whether any important questions were omitted from the naval program of experimentation. They also request an assessment of the adequacy of processes and methods, particularly in planning. As discussed below, these are not unrelated questions. In short, there are key omissions from the current program of experimentation, and some of these result from inadequate methods.

Surveying Future Challenges

A key strategic question is whether the naval experimentation programs are focused well strategically. The committee concludes that, at the highest level, the Navy and the Marine Corps are both doing well here. They are both "capabilities-oriented," as distinct from being oriented toward specific scenarios. This is consistent with DOD guidance[11] and long-standing naval traditions. For example, the components of the new naval force strategy Sea Power 21 (e.g., Sea Strike, Sea Shield, and Sea Basing) are quite general. So also, the primary enabler being emphasized (FORCEnet) describes a broad capability rather than something designed for a narrow function suitable in only special cases.

---

[11]See Donald H. Rumsfeld, Secretary of Defense, 2001, *Quadrennial Defense Review Report*, Washington, D.C., September 30. The classified defense guidance issued in 2002 reportedly reinforced the emphasis.

It is less clear whether this breadth of coverage is realized in more explicit detail. The committee did not see evidence of a comprehensive and rigorous survey of capability needs. The naval experimentation program certainly addresses issues that have been highlighted in high-level planning documents, such as assuring access and projecting force even in the presence of asymmetric strategies. However, the translation of such high-level challenges into concrete problems to be worked on and into priorities for doing so appears to be accomplished in a rather unstructured way, without the benefit of rigorous studies and critiques.[12]

Another concern is that the number of concepts being pursued in experimentation is small and seems to be driven by particular ideas (such as the crosscutting value of network-centric operations) rather than by an attempt to be systematic and to address problems that the Navy knows it is going to face. These would include problems arising either because threats can be foreseen or predicted, or because new capabilities are being pursued and their effects and modus operandi will have to be explored in detail. Arguably, the NWDC and the MCCDC should be responsible for addressing the full range of worries about future naval capabilities, not just one or a few concepts. This is especially so in connection with planning for major transformational changes; history is not encouraging about such matters. Often, nations have worked diligently to develop and perfect concepts that ultimately proved wrongheaded or unsuitable to the war that actually developed.[13]

A rejoinder is that there are not enough resources to do everything at the same time. In fact, however, it *is* possible to address a wide range of problems in parallel. This is an essential part of planning under uncertainty, as is required in capabilities-based planning. To be sure, only a few concepts can be addressed with the highly personnel- and fleet-asset-intensive activities that are usually equated with fleet battle experimentation. However, there is no need to pursue everything using large field experiments. Analytical studies and small-scale experiments would suffice for addressing many issues.

Certain important areas are not yet adequately explored in the naval experimentation programs. Some of these are currently gaining definition.[14] Many

---

[12]The concern here is not just the range of political-military scenarios but also the range of assumed circumstances (e.g., the enemy's strategy and tactics, initial conditions that may include surprise or other setbacks, the nature of both sides' allies, the real-world effectiveness of weapons and systems never used in such circumstances before, and so on).

[13]See Paul K. Davis, 2002, "Integrating Transformation Programs," *Transforming America's Military* (Hans Binnendijk, ed.), National Defense University Press, Washington, D.C., pp. 193-218.

[14]In January 2003, the CNO requested that the Commander, Fleet Forces Command—as part of the lead role for Sea Trial in support of Sea Power 21—"draft and implement a comprehensive roadmap (by May 2003) that integrates studies, wargames, experimentation, and exercises with evaluation metrics and an execution timeline." (Commander, Fleet Forces Command. 2003. *Sea Trial—Concept Development and Experimentation Campaign Plan (U),* Working Paper (draft), Norfolk, Va., May (Classified).)

candidate areas for experimentation emerge from the spectrum of problems facing the Navy in implementing Sea Power 21. Others emerge from questions about capabilities expected to be delivered by programs of record; many of these have been experimented with, but not conclusively, and deserve further investigation. They include over-the-horizon, time-critical strike; use of extended-range guided munitions (ERGMs) for long-distance, high-volume, rapid fire support; expanded applications for network-centric capability to deployable undersea sensor arrays; mine/countermine warfare; the use of UAVs to locate and identify targets; operational concepts for cruise missile submarines; Marine operations in built-up areas; sea-basing concepts such as logistic ship configurations with loadable containers; and advanced support and logistics concepts. The experimental work done to date has brought out overarching issues, such as achieving a satisfactory common operating picture (COP); deconfliction;[15] and bandwidth size and management. And some important areas not yet explored include Vertical Launch System (VLS) reloading at sea, assault breaching of mine and obstacle fields near and on the beach, and continued decisive operations under impaired network conditions and under unfavorable environmental conditions.[16]

The committee believes that these areas, some of which were not mentioned in its Navy and Marine Corps briefings, anticipate legitimate issues with many facets and typify the kinds of experiments that the Navy and the Marine Corps should consider moving from the program of record toward the full Sea Power 21 capabilities. Furthermore, many of these individual areas of experimentation would fit into larger experimentation campaigns, as discussed in that context in Chapter 2. The NWDC recently drafted a Sea Trial experimentation campaign plan.[17] The committee believes that this is a step in the right direction, though the plan's impact on the Navy is as yet unclear. Nevertheless, the noted omissions and inadequacies sustain the committee's concerns about the need for a more robust program and result in a finding for the Navy:

Finding for Navy: As yet, no cohesive experimentation program exists that will move the Navy's forces to "future" concepts, processes, doctrine, and capabilities.

In further assessing Navy and Marine Corps experimentation, other questions arise: (1) Does concept development address the range of challenges? (2) Are *multiple*, competing concepts developed and debated for each? (3) Are

---

[15]Deconfliction is normally understood as "deconfliction of air space" to ensure that two air vehicles (aircraft, missiles, projectiles) do not occupy the same air space at the same time.

[16]An expanded set of candidate areas for future experimentation is provided in Chapter 6.

[17]Commander, Fleet Forces Command. 2003. *Sea Trial—Concept Development and Experimentation Campaign Plan (U),* Working Paper (draft), Norfolk, Va., May (Classified).

the concepts generated well chosen and well conceived? These are considered and answered in the discussion below.

Navy Warfare Development Command

The committee was impressed with how much the Navy Warfare Development Command has accomplished over the past few years. The progress made in a short time is remarkable. One of the laudable, recent changes is that of sharpening the extent to which concepts drive experimentation. That change is likely to improve the coherence of the naval experimentation program as well as to improve the sharpness and quality of the concepts themselves. The command's vigor and enthusiasm were also high, and the command could point to recent successes related to high-speed vessels and the FBE-J experiments. However, as indicated by the committee's discussions and reading, it seemed that the number and range of operational concepts being pursued are small, that debate is informal and internal, and that—despite its efforts—the NWDC has not been sufficiently connected to all of the potential stakeholders (e.g., the fleets and the acquisition community). All of these factors underscore the committee's concerns about the adequacy of the future program of experimentation, already noted in the finding directed to the Navy in the preceding subsection. The CNO's decision to align the NWDC under the Fleet Forces Command may prove useful in mitigating these problems, but may also result in too great a focus on the near term.

The committee was troubled by the sometimes overly rapid jump from "concept development" (which is sometimes criticized as too "soft" or "conceptual," without sufficient grounding in engineering and other technical analysis) and longer-term commitment to high-stakes experimentation and even fast-track acquisition. A number of those consulted during the course of this study acknowledged that this was a problem. Unfortunately, the Navy's acquisition community, which has the knowledge and expertise to translate soft concepts into more rigorous constructs and capability needs, is sometimes only minimally involved, and often not at all.[18]

One reason that the acquisition community is not more closely involved on a routine basis may be leadership interest in using experimentation to bypass the sometimes-ponderous acquisition system. The desire to speed up innovation is both understandable and laudable. However, there is clear need for a process in which concepts are exposed to initial systems analysis and systems engineering before they are tested in the field and during the field evaluation process. This step, which does not exist now, is needed and could be accomplished by involv-

[18]Arguably, the problem is process—the lack of a clean *mechanism* for transitioning from successful experimentation to acquisition. Individuals sometimes transcend the system, but success depends on personal relationships and efforts that surmount ongoing processes.

ing both stakeholders and well-chosen teams from the relevant Systems Commands in carefully crafted (but not necessarily large) experimentation events.

Marine Corps Combat Development Command

Currently, the Marine Corps appears to be operating effectively in its concept formulation—in part because of the nature of the concepts it is exploring and in part because it operates in a tightly integrated system. The organizational structure of the Marine Corps, with the MCCDC responsible to the Commandant for experimentation and also responsible for DOTMLPF integration across all capabilities, facilitates a thorough examination of all results from several perspectives. Requiring that the Marine Requirements Oversight Council (MROC) approve the Marine Corps experimentation plan and review all claimed results helps to keep efforts focused in the intended direction, while causing senior leadership—not just the Commandant—to take responsibility and to share ownership. Also, MROC involvement helps create a culture that is proactive toward experimentation. All of the Marine Corps Service experiments are linked directly to advanced warfighting concepts. The experiment objectives and concept linkage are published in the respective experiment plans.

One major drawback of the current approach is that the Marine Corps experimentation plans are not focused on long-term capabilities for the forces. Instead, experimentation plans and objectives are tied to change and improvement, tending to deal with small-scale issues and to address improved infantry TTPs. They are along the lines of moderately paced evolution consistent with the intuition of current senior leaders. Experimentation needs to address challenges and capability improvements for the near, middle, and long term, although the respective levels of effort can and should vary substantially. For balance, there should be a stronger component of more revolutionary experimentation.[19] The preceding observations lead to this finding:

Finding for Marine Corps: The Marine Corps has moved from a balanced program of experimentation campaigns based on near-term, mid-term, and long-term objectives to one of experiments focused on near- and mid-term objectives.

The committee observes that many problems of a long-term nature need continuing attention in experimentation, sustained over time so as to affect capabilities in the field. Examples include the following:

---

[19]Early aspects of the Sea Dragon exercise program had this character, but it proved difficult to move from experimentation into fielded capability. The current approach is more evolutionary but also more effective organizationally in bringing about continual change.

- *Sea basing.* This concept establishes a commendable new approach for the operation of naval forces. It requires only minimal dependence on the assistance of host nations and their facilities. Close cooperation with the Navy experimentation community and others is required to continue exploration and development of the sea-basing concept. Leaving much of the force's support at sea will require new designs for ship or other types of offshore platforms that allow selective offloading and reconfiguration of cargo containers on board; at least limited on-board support of large, vertical-lift aircraft, if not more radical solutions; lighterage for over-the-beach logistics in rough sea conditions up to sea-state 3 or 4; and the ability to support forces ashore from many miles out at sea in order to avoid shoreward defenses such as mines and short-range antiship missiles. These are radically new conditions for supporting a naval force ashore. Learning how to implement them effectively will require much experimentation, from the level of the individual ship to that of large-scale forces. Such a program of experimentation remains to be designed, funded, and carried out. And once a problem's physical parameters have been ascertained by such means, the results will have to be translated into new systems, doctrine, and procedures that will have to enter the naval forces' budgets, acquisition, training, and operational programs according to schedules designed to bring the capabilities into effect as early as feasible.
- *Littoral combat.* U.S. naval power projection is built on the foundation of mastery of the seas—being able to operate where and when required. Today, this mastery exists on the open seas—on "blue water"—but it is not fully realized for "brown water" littoral naval forces. Major problems in littoral waters include enemy resistance in the form of mine warfare, antiship missiles, armed small boats that can attack and damage U.S. logistics ships, and possible use of chemical or biological weapons against near-shore operations of naval forces. In addition, the complex orchestration of Marine Corps landing and support forces, Navy surface and air fire support, and defensive operations to protect the landing forces at sea and ashore will require many changes in the C4ISR system on a joint Navy and Marine Corps basis. These changes will also have to be extended to other Services that may be involved in an operation. Much experimentation remains to be done, from the small-unit to the force level, to devise the system requirements, the TTPs, and the joint C4ISR coordination for full success in such operations.
- *Unconventional warfare.* Unconventional warfare, as exemplified by the operations of Marine Expeditionary Units (Special Operations Capable), has been part of naval operating procedures for many years. However, changes in sea basing and brown water littoral operations will require the adaptation of current doctrine and TTPs. Experimentation under these new conditions or as part of support to the Special Operations Forces-related naval forces will be necessary. For instance, with more use of unmanned underwater vehicles, unmanned aerial vehicles, and new modes of landing, it will be necessary to provide fire power

and logistics support, commercial transport, pathfinder, and the Seals' underwater vehicle operations.

Methods—Experimentation on Concepts

Assuming that good concepts and their priorities are established, the next question is whether appropriate methodology is employed to pursue the concepts that undergo systematic experimentation. Chapter 2 describes a canonical approach to experimentation campaigns, which involves an overall campaign strategy, warfighter reviews, wargaming, simulation, and field experiments (as are needed and feasible). Studies and analyses constitute a systematic, overarching activity, rather than merely a preparation for and analysis of data from some particular experiment. Analysis should encompass the development of theory to make sense of the whole, systems analysis and systems engineering, empirical analysis, and, ultimately, policy analysis leading to choices among the concepts regarding their embodiment in doctrine and their further development for fielding.

During the past few years, the Navy, the Marine Corps, other Services, and the U.S. Joint Forces Command (USJFCOM) have all embraced this approach to some degree. Those involved in experimentation programs usually acknowledge the importance of all of the activities (not just the big events). The NWDC and the MCCDC both construct experimentation campaign plans with a mix of the various types of venues and tools. Both organizations also work with USJFCOM to develop the USJFCOM campaign plans mandated by Congress.[20] All of these campaign plans reflect considerable learning over the last 6 years or so.

Despite this progress, the committee concludes that preparing for the event of a fleet battle experiment has largely (but not exclusively) been the focus of attention in the Navy's experimentation program. Studies, analysis, and "writing the book" (i.e., developing definitive knowledge and understanding aimed at applying the results) should play more central roles, but they are underemphasized activities today. The NWDC's experimentation campaigns require expansion and structure to ensure thorough learning of the relevant phenomena and options for improved capabilities by careful selection of experimentation venues, rather than overemphasis and over-reliance on the success of a major event. This is a relevant and needed change in strategy for the Sea Trial process.

The "big event" of a fleet battle experiment is very important and sometimes critical for exploring scalability and integration, demonstrating potential, and building broad-based enthusiasm and support. But it is neither an end in itself nor, in many cases, an appropriate culmination. After such an event, more remains

[20]See, for example, U.S. Joint Forces Command, 2000, 2001, and 2002, *Joint Concept Development and Experimentation Campaign Plan*, Norfolk, Va., September. Chapter 4 discusses USJFCOM developments in more detail.

to be learned, partly from operator feedback, partly from continued experimentation. Furthermore, the circumstances of the big event are often singular and atypical of wartime operations in at least some important respects, such as force structure and threat scenarios.

An interesting contrast is struck between the classic development of nuclear submarines and sea-launched ballistic missiles on the one hand,[21] and the current pursuit of FORCEnet, on the other. The former was legendary for its depth, quality, discipline, and engineering prowess. The latter, as best the committee could judge from presentations and materials received, is being pursued in an unstructured way, including relatively ad hoc experimentation with little underlying architecture and design. Today discipline in developing and structuring campaigns is sometimes evident, but oftentimes not.[22] The emphasis appears to be on concepts that are sometimes not fully developed, competed, and critiqued and their demonstration under circumstances that do not provide the information required.

The NWDC has made some changes in processes. In the past, concepts were proposed at a high level, building momentum that would extend throughout the experiment planning cycle. The maturation and development of a concept could extend up to the final planning conference for the experiment. Initiatives, derived from concepts, were also kept at a high level, and the objectives for experimentation then followed; these too could be introduced late in the process. Such continuous changes in concepts and objectives placed pressure on the experiment design.

In a recent approach to planning, concepts were developed by Warfare Innovation Development Teams (WIDTs).[23] As concepts and doctrine co-evolved, a set of related objectives were developed for the FBE. Concepts and associated objectives were produced at the front end of the experimentation process and were more stable throughout that process. Each team was responsible for a set of activities that consist of war games, meetings, limited-objective experiments (LOEs), exercises, real-world experiences, and so on. FBEs may or may not be the culminating event.

---

[21]In Chapter 3, see the subsections entitled "A Case History in Past Experimentation: The Early Development of Naval Aviation," and "Evolution of the Linear Development Model" for a discussion of past successes in developing new capabilities.

[22]Planning for Fleet Battle Exercise-Juliet (FBE-J), which was conducted in 2002, was much more systematic and rigorous than planning for the previous FBEs. However, the focus was on methodology for the big-event experiment, not on preparing a definitive analysis of, say, naval firepower in the network-centric era.

[23]The committee understands that NWDC has recently changed this process somewhat. There are still five WIDTs, now called Sea Strike, Sea Basing, Sea Shield, Information and Warfare Advantage, and Combating Terrorism/Force Protection. The committee believes its comments are applicable to this process even if organized differently.

The committee believes that these changes hold promise for some positive consequences. They improve stability for good experiment design and place emphasis on selecting the right kind of activity to address the problem at hand. Nevertheless, as indicated in Chapter 3, there are concerns with this process:

- The process by which a concept becomes worthy of further exploration by a WIDT is not clear.
- Although more time may be available for better experiment design and planning, the process must be stable enough to produce better results.
- It is also not clear how feedback and knowledge gained from other events and venues contribute to the definition of new concepts and experimentation objectives, and result in coherent experimentation campaigns.
- While the importance of campaigns was acknowledged, no examples of how the process results in experimentation campaigns were forthcoming.
- Participation in the WIDTs was still primarily an internal function of NWDC, without much participation from outside stakeholders.
- Analysis and objectives were still emphasized at the end of the experimentation process, rather than at the beginning.
- Cross-concept analysis was further fractured, as there was not a clear relationship between learning from one WIDT to another, or in the development of experiments.

The assessment of experimentation methods leads to another finding for the Navy.

Finding for Navy: There are significant deficiencies in the end-to-end processes of naval experimentation. Shortfalls include the following:

- Insufficient use of structured experimentation campaigns;
- Inadequate breadth in exploration of concepts;
- Insufficient studies and analysis, including their use to determine the best experimentation venue (games; simulations; small, focused experiments; and so on);
- Inadequate use of spiral processes to build knowledge iteratively; and
- Inadequate planning and evaluation.

The Influence of Methods on Resources

The committee acknowledges the value of FBEs, both for supporting various objectives and for gaining the enthusiasm and support of influential senior leaders (such as fleet commanders) by making tangible what had previously been abstract and hypothetical. The history of military innovation is replete with examples of analogous demonstrations.[24] The big experiments also reflect long-standing traditions of "just trying things." However inefficient scientifically, such experiments bring warfighters into the process of change, rather than merely allowing work to be done in laboratories and other R&D venues in the hope that it will prove interesting to warfighters eventually. And, finally, there have been quite a number of significant results from fleet battle experiments.[25]

Nonetheless, the focus on big fleet experiments comes with a significant price tag, typically in the few millions of dollars, though costs were much higher for Navy participation in the recently conducted Millennium Challenge '02 (MC 02) joint experiment. There are also substantial opportunity costs in terms of the time of required key individuals and critical fleet assets, not just funds. According to some participants the prevailing incentive structure works against systematic effort because the expectation is that large experiments must not fail in the achievement of their objectives, however much it is argued otherwise. Thus, these experiments are often constrained to prevent the range of exploration needed. Even sheer size contributes its own diminution of free play, as noted in Chapter 4. Depending on how the program of joint experiments managed by USJFCOM develops, the problem could worsen, since the process of staffing and otherwise preparing for a major joint exercise/experiment such as MC 02 is all-consuming.

In recent years, FBEs have become progressively more complex and have incorporated more tests and experiments within a limited period of platform and asset availability. As a consequence, it can be argued that the experiments undertaken during FBEs are inherently less complete than those undertaken during LOEs. For example, the operational parameters and attributes of the HSV could not possibly have been completely explored during the few-week period associated with an FBE. An LOE dedicated to an examination of the attributes of the HSV was used in addition to its participation in a FBE. As a result, the Navy has a much broader understanding of the HSV than it would have gained in the course of an FBE. The committee's conclusion is that the overemphasis on FBEs is not always the best use of the limited resources available to experimentation.

Another by-product of the emphasis on FBEs is the compromise or modification of objectives that results from combining experiments with exercises and maintenance schedules. These activities have, by nature, inherently different pur-

---

[24]See, for example, Andrew F. Krepenivich, 2002, *Lighting the Path Ahead: Field Exercises and Transformation*, Center for Strategic and Budgetary Assessment, Washington, D.C.; and Richard O. Hundley, 1999, *Past Revolutions, Future Transformations: What Can the History of Revolutions in Military Affairs Tell Us About Transforming the U.S. Military?*, RAND, Santa Monica, Calif.

[25]In Chapter 3, see the section entitled "Synopsis of Results to Date from Fleet Battle Experiments Alpha Through India.

poses and contexts. Exercises, for instance, address readiness and are associated with current doctrine, organizations, and immediate needs. Experimentation deals with evolving DOTMLPF. Owing to the need to align FBEs with the availability of assets such as ship platforms and aerial assets, many experimentation objectives must be substantially modified or jettisoned. Those that remain are examined in a field environment that is not always suited for proper data compilation methods or that does not always readily factor in anomalies and artificialities introduced during the conduct of the experiment. While good experimentation design should temper such effects, these circumstances do often limit the validity of the results—underscoring the concern that the FBE is not always the most effective use of limited resources.

Funding for an FBE is provided by the NWDC, which also must use its coordinating skills to leverage participation, since it does not have the line-item funding to control equipment used for experimentation. Other organizations, such as the Office of Naval Research (ONR), provide support for the NWDC's planning of FBEs as well as sponsoring individual projects and concepts that are evaluated in an FBE. ONR has nominally intended to provide from $20 million to $40 million per year for experimentation, but in the past, funds have been used for other higher-priority efforts. Such diversion of funds has occurred with other sponsors as well. These perturbations can have severe impacts on the planning, preparation, conduct, and evaluation of FBEs, which typically have long cycle times of 12 to 18 months. This situation is conducive to a poor return on investment, resulting in delays and/or a loss of substantial effort, with high potential for the compromise of objectives.

While the committee deliberated about making recommendations for increased funding for experimentation, changes in already-programmed funding were sufficiently frequent and severe as to make the need for and scope of such recommendations unclear. What is clear is that the Navy is not making effective use of the experimentation resources already programmed. Also clear is a need for the Navy to establish a line item for funding experimentation that reflects the value placed on experimentation, stands on its own merits, and is supported and sustained Navy-wide. As a notable aside, the submarine community's development squadron (DEVRON 12) organization, well regarded for its experimentation successes, controls its own assets and budget for experimentation, in contrast to the situation at the NWDC.

These observations result in an additional finding.

Finding for Navy: The Navy has not made effective use of resources in its experimentation program.

A rebalancing of resource investment requires an increased emphasis on analyses, studies, games, models and simulations, and limited-objective experiments of various types, leading up to field experiments whose scale is carefully and analytically moderated.

Although the Marine Corps emphasizes field experiments of various types (in large part because it is difficult to evaluate infantry concepts in other ways), the concepts are subjected early to a series of formal and informal war games, tabletop discussions by subject-matter experts, seminars, symposiums, and, if possible, modeling and simulation (M&S) prior to field experimentation. Some limited experiments lend themselves well to M&S. The Marine Corps projects Einstein and Isaac[26] are efforts that depend heavily on M&S; the results have been folded into major experiments, notably Urban Warrior.

The situation in the Marine Corps has not always been so balanced. Only a few years ago, the Corps was overly reliant on large experiments—even for the purpose of learning lessons that could have been obtained with smaller experiments. Also, as discussed in Chapter 3, early work in the Sea Dragon series was not always well planned, and preparation for experiments was not always adequate. Consequently, some equipment problems occurred that could have been anticipated, mitigated, or avoided. However, the Marine Corps learned and altered its approach to experimentation.

One change that it has made is more systematic preparation for field experiments (e.g., scheduling time to train personnel on new equipment and in the new concepts and procedures before experimentation begins). Another has been and continues to be a growing reliance on smaller experiments (including limited technical assessments, or LTAs). The committee found these changes and trends on the part of the Marine Corps encouraging.

ASSESSMENT OF ENVIRONMENT, INFRASTRUCTURE, AND TOOLS FOR EXPERIMENTATION

Deriving the maximum benefit from experimentation requires a suitable environment, as discussed in Chapter 2.[27] One of the specific questions from the terms of reference is how adequate the tools and environments for experimentation are (e.g., the Navy's modeling and simulation capabilities, integration facilities, and so on). This section answers this question and examines the naval environment, infrastructure, and key tools for experimentation. The results of the assessment are summarized in Table 5.2 and the detailed discussions leading to these conclusions are presented in the subsections that follow.

---

[26]See the Web site of Andrew Ilachinski, Center for Naval Analyses, at <www.cna.org/isaac>. Accessed October 9, 2003.

[27]See the section entitled "Environment for Experimentation."

TABLE 5.2 Assessment of Navy and Marine Corps Environment and Infrastructure

Enabler	U.S. Navy	U.S. Marine Corps
Leadership and culture • Committed leaders • Learning culture • Incentives for risk • Tolerance of negative results • Empowerment	Strong culture of innovation and adaptation to challenge. Top-down leadership is needed to provide minimal coherence for an experimentation program. Innovators are not visibly rewarded in the Navy Warfare Development Command (NWDC) experimentation programs. Concept developers are at the end of their careers.	Strong culture of innovation and adaptation to challenge. Risk taking and horizon of inquiry have varied. May need to spin off and protect a subset of more radical experiments.
Trained and talented people • Concept developers • Systems analysts • Operators • Red-team cells • Support teams	Shortages in higher-level skills, especially systems analysts, systems engineers, and architects. Potential decommissioning of Third Fleet's USS *Coronado* will reduce needed experimentation expertise. Naval Postgraduate School experimentation expertise may be disbanded.	Shortages in higher-level skills, especially systems analysts, systems engineers, and architects.
Information and physical infrastructures • Networks • Information repositories • Architectural frameworks • Integration and test facilities • Training facilities • Places and platforms	Great opportunities ahead for fleet-centered embedded and distributed exercise/experimental command and control (including reachback); NWDC and Marine Corps Combat Development Command need to track actual progress. Next-generation mission scenarios are needed. Necessary ships and airborne command, control, communications, computers, intelligence, surveillance, and reconnaissance (C4ISR) assets are often unavailable. USS *Coronado* is an exceedingly valuable asset but may be decommissioned. Distributed engineering plant may be useful for spiral processing in experimentation.	Great opportunities ahead; need to track actual progress in improving network-centric environment in the fleet. Next-generation mission scenarios are needed. Infrastructure good and improving, e.g., new command and control facility at Camp Pendleton, California (Marine Corps Tactical Systems Support Activity).

continues

TABLE 5.2 Continued

Enabler	U.S. Navy	U.S. Marine Corps
Tools • Modeling and simulation • Prototypes, surrogates, and so on • Artificial environments • Data capture and dissemination	Simulations have supported major field experiments but are not adequate for representing future warfighting concepts. Focus has been on simulations to support live forces. Significant benefit in working with the U.S. Joint Forces Command (USJFCOM) to develop future simulation environments. Greater emphasis on supporting the end-to-end range of experimentation activities through modeling and gaming capabilities and virtual simulation environments is necessary. Data collection and analysis tools are also needed.	Simulations have supported major field experiments but are not adequate for representing future warfighting concepts. Significant benefit to working with USJFCOM to develop future simulation environments. Data collection and analysis tools are also needed.

Leadership and Culture

The committee concludes that both the Navy and the Marine Corps have strong, dynamic cultures that believe in adaptation and innovation. Furthermore, both Services have used experimentation as an enabler of change for decades.

Navy Leadership

The current Chief of Naval Operations (CNO) is relying on experimentation through Sea Trial initiatives to enable his Capstone Concept Sea Power 21. The connection that he makes between the use of experimentation and the development of such advanced concepts as FORCEnet raises the bar on his expectations for experimentation and thereby promotes greater initiative for innovation. The CNO also realigned the NWDC under the Fleet Forces Command so that it would be tied more closely to the fleet, reversing his predecessor's decision on the matter. These actions illustrate the CNO's interest in experimentation, his belief in its relevance to the fleet, and his continuing support for it, including the use of fleet battle experiments. However, the committee has substantial concerns about the lack of an alignment of the experimentation organization (NWDC) and the experimentation program (Sea Trial) directly under the CNO. One lesson learned from past success in experimentation suggests the need for this strong a connection to the most senior leadership. This is particularly pertinent when experimentation is intended to result in significant new capabilities for the field.[28]

Certainly the realignment of the NWDC enhances the authority of its commander, enabling him to garner more support and provide increased coherence for the experimentation program, though the committee acknowledges that this move under the Fleet Forces Command is relatively recent. In working with the many Navy organizations participating in experimentation, the NWDC needs to use its coordinating skills and abilities to direct and leverage all of the participation it can obtain from organizations and parties that it does not control. The NWDC does not have line-item funding to control equipment to be used for experimentation (for example, it relies on ONR), nor can it require the N7 staff to act on the results of experiments when dealing with Navy "requirements."

The lack of the direct connection from the NWDC to the CNO, together with another factor, compounds the committee's concerns: that is, a potential by-product of the realignment is an overemphasis on the near-term needs of the fleet. The likelihood of this was viewed as high by members of the committee because there will be continuing tensions between experimentation and readiness-related activities. These concerns result in the following finding:

Finding for Navy: Continuing Chief of Naval Operations and Vice Chief of Naval Operations sponsorship, leadership, and attention are vital for the overall coordination, direction, prioritization, and execution of the naval experimentation program.

Navy Culture

The Navy's incentive structure is not uniformly good within the experimentation community.[29] In particular, opportunities for promotion from or extended tours in the NWDC do not appear to be realized. Indeed, many officers there appear to be in their last tour. Further, it is not evident how much influence and prestige the position of NWDC's commander is intended to have. It is hoped that the realignment under the Commander, Fleet Forces Command (CFFC), and the promulgation of Sea Trial will increase these aspects of the position. A contrast is provided by the submarine community with the well-regarded, dedicated submarine development squadron (DEVRON 12), which enjoys high prestige as it

---

[28]As evidenced by examples discussed earlier in this report—naval aviation, ARCI, Army digitization, and the Air Force Command and Control Center.

[29]Other innovators in the Navy are rewarded. For example, members of the CNO's Strategic Studies Group at Newport, Rhode Island, are regarded as prime candidates for promotion and future leadership positions. They also work on concept development but may or may not be regarded as part of the experimentation community.

pursues continuous experimentation and improvement. The new commander of the U.S. Joint Forces Command is one alumnus of that activity.

Marine Corps Leadership

The Commandant of the Marine Corps has also taken measures to improve the effectiveness of experimentation. He has created the MROC, which, in addition to its other responsibilities, provides early and continuing attention to experimentation by the most senior commanders. This also makes experimentation more institutional and less dependent on the Commandant. By implementing process changes in a series of MROC meetings, the Assistant Commandant has also provided strong guidance and direction to wed the Service experimentation to that of the joint community.

Marine Corps Culture

One questionable change within the Marine Corps in recent years has been increased emphasis on near-term and mid-term objectives in experimentation at the cost of long-term objectives, noted in a prior finding. This change reduces the prospects for risk taking in experimentation, whereas more risk taking is needed to achieve objectives that are transformational. Longer-term change traditionally depends on pursuing highly disruptive technologies and concepts, and the people advocating such departures, even through experimentation, are themselves often viewed as disruptive. There is a need to encourage and reward those who advocate high-risk possibilities, while protecting and balancing innovative but high-risk experimentation.

Despite the need just noted, incentive structures appear to be better for individuals associated with experimentation in the Marine Corps than in the Navy. This may be due to the tight-loop connection with the Commandant and other senior commanders.

Trained and Talented People

Both Naval Services have an abundance of high-quality personnel. The Navy lacks a cadre of experimentation experts for in-depth studies connected with experimentation (most such experts could be civilians). Systematic and rigorous experimentation is a complex and technical undertaking that requires dedicated expertise.[30] Both the NWDC and the Third Fleet have dedicated personnel for

---

[30]This can be seen from perusing illustrative reports on experimentation. See, for example, Joint Advanced Warfighting Program of the Institute for Defense Analyses, 2000, *The Joint Experiment J9901: Attack Operations Against Critical Mobile Targets*, Institute for Defense Analyses, Arlington, Va. (limited distribution). The committee also saw good examples of this in discussions with

concept formulation, some experiment planning, and experiments themselves. Nonetheless, experimentation programs of the Navy and the Marine Corps have shortages of personnel with certain high-level skills, notably, systems analysts, systems engineers, and systems architects.

A cadre for in-depth work in experimental design had been constructed at the U.S. Naval Postgraduate School for Fleet Battle Experiment-Juliet, but it may be disbanded in the near future—apparently because its activities were not seen as fitting in to the school's mission. Also, at the time of this writing there was a movement by the Navy to decommission the USS *Coronado*, which would impact the Third Fleet's personnel support.[31] One option that has been mentioned is reducing the *Coronado*'s status to that of a reserve ship, which could reduce yearly costs by perhaps $10 million to $20 million since it would no longer be required to have the same level of readiness as that of an active-duty command ship. That might be a reasonable alternative, but the committee was concerned that such a change would reduce the prestige and career-enhancing aspects of being assigned to experimentation—as well as removing a useful platform.

Recent activities are also having an effect on the numbers of Marine personnel engaged in experimentation. The Service is experiencing increased competition for its personnel assets, primarily from the demands engendered by large-scale joint experimentation and also by the level of participation required by the new joint processes. The Navy is also affected, though less so owing to its size. The press on the Marine Corps staff has been especially severe, since its resources are small and the priorities of the joint process have at times been duplicative and/or conflicting and therefore inordinately demanding, although the latter difficulty may be being mitigated.[32] The committee views these impacts as negative, since they result in fewer resources available for other activities so important to good experimentation, such as analyses. The committee also notes this as another example of the strain that can be put on available personnel by large experimentation events.

Information and Physical Infrastructure

The committee believes by and large that the quality of the information infrastructure is improving, both for naval experimentation and in the forces

---

Dr. Shelly Gallup of the Naval Postgraduate School. See also David S. Alberts, Richard E. Hayes, John E. Kirzl, Leedom K. Dennis, and Daniel T. Maxwell, 2002, *Code of Best Practice for Experimentation*, DOD Command and Control Research Program, Office of the Assistant Secretary of Defense (Networks and Information Integration), Washington, D.C., July, Ch. 6. Available online at <http://www.dodccrp.org/>. Accessed October 7, 2003.

[31]The USS *Coronado* is slated for decommissioning in FY 2005.

[32]In Chapter 4, see the subsection entitled "The Role of Joint Experimentation in Preparing for Joint Operations."

themselves. On the basis of recent experiments and infrastructure developments at the Navy's Distributed Engineering Plant and the Marine Corps Tactical Systems Support Activity (MCTSSA), discussed below, the networking environment appears adequate for purposes of conducting the experiments. Significant shortcomings exist in the network for supporting operational missions (and hence operational missions as examined in experiments), but that is a different matter from the experimentation infrastructure. Nonetheless, the adequacy of the experimentation information infrastructure should be reassessed as network-centric experimentation moves forward and as key concepts under Sea Trial are explored. Experimentation with network-centric operations may create tremendous but as yet undefined challenges that will have to be accommodated through additional infrastructure and tools support.

The Navy has invested successfully in the Distributed Engineering Plant (DEP)[33] for the resolution of critical fleet interoperability issues. These are issues spawned by the growth in tactical networking capabilities such as Link-11, Link-16, and the cooperative engagement capability (CEC). While the main mission of the DEP has been the testing and certification of battle groups before they deploy, it is providing support in system design and development, prototyping for the CEC, and validation of operational processes and TTPs. Consequently, it may be a potential source of infrastructure for experimentation, particularly in spiral development. Its adequacy to support tool interoperability and integration should be assessed as future experimentation campaigns are developed.

Navy

Currently the Navy lacks some key platforms for experimentation. The submarine community usually has adequate access to required platforms for experimentation because it alone controls those platforms. However, there are some problems related to the availability (and budget) for experimentation with ships and C4ISR assets and with manned and unmanned aerial systems such as the U-2 and Global Hawk that are owned by the other Services. The Third Fleet's USS *Coronado* is valuable because it does provide a dedicated command-ship platform, but, as noted earlier, this asset is currently scheduled for decommissioning in the interest of cost cutting.

No clear solution is apparent for having access as needed to aerial and non-Naval Service C4ISR assets, but their importance is growing and the experimen-

---

[33]The DEP is a founded on shore-based sites that replicate the hardware, software, connectivity, and environment of the ship and combat systems and interconnect them to replicate a battle group. See Jeffrey H. McConnell, Technical Coordinator, Navy Distributed Engineering Plant, Naval Surface Warfare Center, Dahlgren, 2002, "The Navy Distributed Engineering Plant—Supporting Force Systems Engineering," viewgraphs available online at <http://www.dtic.mil/ndia/2002systems/mcconnell2a4.pdf>. Accessed October 9, 2003.

tation program needs access to these systems. The NWDC's new position under CFFC may improve prospects in this area.

Marine Corps

The Marine Corps appears to have adequate physical infrastructure (especially with the addition of MCTSSA, the new facility at Camp Pendleton for command and control), but with the current and projected growth of C4ISR systems, additional options may eventually be required. Also, the Marine Corps has the same problem as the Navy with respect to access to some C4ISR and aerial assets that are not owned by the Service.

The other Services and the U.S. Joint Forces Command are increasing their investments in linked computer simulations and ranges, as well as in artificial and virtual environments. As standards for interoperability among facilities and ranges are developed, the Marine Corps may need to invest in these areas to offset the projected rise in operations and personnel tempo for people and equipment. These investments would not only permit both training and experimenting in virtual, realistic environments but would also reduce costs associated with equipment transportation, maintenance, and duplication of excess personnel and equipment (hedging against contingencies), as well as personnel transportation.

A relevant example is that of the Army's long-term investment in systems with which to train personnel for operations similar to that in Bosnia. One effort, at the University of Southern California, is researching how to have a trainee (e.g., a lieutenant) interact with a virtual-reality "world." This world has relatively realistic, animated, artificial-intelligence beings representing, for example, villagers and soldiers at an intersection where a traffic accident involving U.S. forces has occurred. The trainee must assess the situation and interact with the "people" (who respond to the lieutenant's voice, show emotion, ask questions, and take actions).[34]

Tools

Tools for Analytical Work

It was clear from presentations and documentation provided to the committee, as well as from discussions with presenters, that the Navy's analytical work

---

[34]See William Swartout, 2002, "Creating Human-Oriented Simulation: The Challenge of the Holodeck," presented at the Grand Challenges for Modeling and Simulation Seminar at Schloss Dagstuhl International Conference and Research Center for Computer Science, Wadern, Germany, August 26-30. Available online at <http://www.informatik.uni-rostock.de/~lin/GC/report/Swartout.html>. Accessed October 9, 2003.

underlying the experimentation program is currently unbalanced in favor of relatively detailed models and simulations tied to the large experiments. Its analytical work is not well suited to broad and rigorous capabilities-based planning,[35] as distinct from working through particular scenarios in more detail. Consequently, there is a shortfall in families of models and games.

Figure 5.1 elaborates on why families of models and games approaches are needed by showing that different members have different strengths. Low-resolution models, if designed for the purpose, can be excellent for exploratory analysis and design—covering a breadth of cases, but not in much detail. In contrast, entity-level simulation can provide a much richer depiction of some underlying cause-effect relationships, i.e., of the underlying phenomena. However, such detailed simulations are not appropriate for broad, design-level exploration,[36] nor even for doing a good job in FBE-style experimentation.[37] Nor are they currently detailed and accurate enough to represent some important issues (e.g., performance of U-2-based C4ISR as a function of operational circumstances, pilots, weather, and so on), which need to be observed with real platforms.[38] War games can bring in human warfighters, who are essential in some play and represent potential users. Field tests with live forces, supplemented by simulations of actual battle, can provide the ultimate experimentation—short of war itself—but are inherently limited in many respects. And, again, they are not very good at supporting understanding of the breadth of possibilities for drawing generalizations, both of which require more abstracted modeling and analysis.

Improving the balance of analytical work underlying the experimentation program would require a realignment of effort within some of the analytical

---

[35]Capabilities-based planning has been mandated since the 2001 Quadrennial Defense Review. For high-level technical discussion, see Paul K. Davis, 2002, *Analytic Architecture for Capabilities-Based Planning, Mission-System Analysis, and Transformation*, RAND, Santa Monica, Calif.

[36]Important, less-abstracted exploration can be accomplished with more detailed human-machine simulations and could be accomplished by real-world operators in the fleet if the tools for doing so were adequately embedded in operational systems, permitting continuous experimentation along with training. This type of exploration is especially suitable for exploiting concepts and technology in near-term improvements. Higher-level exploration—for example, across scenarios and case spaces—is different in character.

[37]For an interesting discussion by one of the developers of the well-respected Naval Simulation System, see William Stevens, 2000, "Use of Modeling and Simulation (M&S) in Support of the Assessment of Information Technology (IT) and Network Centric Warfare (NCW) Systems and Concepts," 5th International Symposium on Command and Control Research and Technology (ICCRT), held at Australia War Memorial, Canberra, Australia, October 24-26, and sponsored by DOD Command and Control Research Program (CCRP), Office of the Assistant Secretary of Defense (Network and Information Integration), Washington, D.C., and Australian Department of Defence, Defence Science and Technology Organisation.

[38]This important point contradicts the notion that simulations can do nearly everything. It was stressed in discussions with Shelly Gallup of the U.S. Naval Postgraduate School, on the basis of extensive experience in fleet battle experiments.

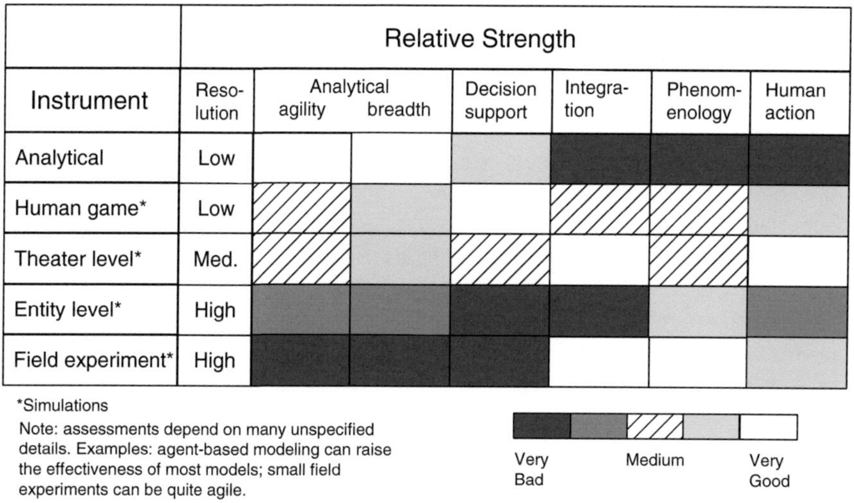

FIGURE 5.1 Comparative strengths within a family of models and games.

groups supporting experimentation, or the creation of new groups. For example, the NWDC's modeling and simulation group has made enormous strides in recent years in establishing the capability to simulate many aspects of operations sufficiently well so that live and simulated play can be intermixed and so that live players can be stimulated by and can interact with simulations in much the same way as with real-world command and control systems.[39] However, these efforts are extraordinarily demanding, in terms of both energy and resources. As a result, the group acknowledges that it does very little of the lower-resolution analysis associated with exploration or systems analysis. One way or another, the Navy needs to supplement these capabilities with other skills (some of which exist elsewhere within the Navy community).

The large and complex simulation work is essential if experimentation is to connect and resonate well with fleet operations and if transitions are to occur easily; it is also essential in many instances in which it provides analytical insights not achievable in other ways (at least currently). Elsewhere this chapter discusses the need to emphasize analytical work and experimentation activities that offer good and correct alternatives to large scale events. The issue is one of balance.

[39]Discussions with Guy Purser of the NWDC and Annette Ratzenberger of USJFCOM.

Tools for Simulation

Properly applied, modeling and simulation can both accelerate the pace of experimentation and reduce its cost. Consequently, it is important to both Service-unique and joint experimentation. There is a need and an opportunity to develop naval use of virtual environments (no interaction with field forces) and constructive environments (some field involvement) for experimentation purposes. As the Navy works through the concepts required for Sea Power 21, virtual and constructive simulations can be extremely useful for developing, exploring, and testing the concepts before trying them out with actual forces.

In assessing the need to expand naval simulation infrastructure for Sea Trial as well as that for future joint experimentation,[40] specific shortcomings in the naval simulation environment emerge. Most simulations exercised by the Navy in the past have involved some forces and have been structured within the context of large field experiments. However, there is a need now for a more robust set of M&S capabilities to align with the many activities involved in the Sea Trial and joint experimentation campaigns, beginning with exploratory concept development.

After MC 02, USJFCOM noted that the DOD's existing M&S capabilities were inadequate to represent future operational concepts. These capabilities do not account for information operations, model new organizations, or capture asymmetric warfare strategies. Also observed after MC 02 was that data analysis tools were too limited and were inadequate for dealing with the large amounts of data collected in MC 02.

In response to these shortfalls in the joint infrastructure, USJFCOM will be addressing an entire range of enhancements through its continuous experimentation environment to enable more flexible, higher-fidelity M&S and through the Joint Simulation System training tool to provide realistic, large-scale, simulations placing humans in operational environments. The development and use of such a common simulation environment are necessary as well as desirable, although the Navy and the Marine Corps will also need their own naval-specific simulations. Such an interoperable environment brings twofold benefits: (1) a consistent environment for both joint and naval-only play and (2) the cost economies of developing one environment to serve two (or more) communities.

The key observations of the committee's assessment of naval simulation environments are based on future needs for Service-unique and joint experimentation. They are as follows:

- While the current modeling and simulation environments have successfully supported major field experiments, there are deficiencies in M&S capabilities

[40]In Chapter 4, see the subsections entitled "Joint Forces Command's Emerging Modeling and Simulation Infrastructure" and "Naval and Joint Linkages in Simulation."

to support a full range of experimentation activities, such as concept development, games, and small experiments; in the areas of compatibility across joint and Service simulations; and in the representation of future warfighting environments.

- The greater use of simulation relative to live forces would allow more experimentation in the future, especially for concepts that apply above the tactical level. In particular, the Navy should make greater use of virtual and constructive simulations for developing, exploring, and testing concepts of operation before trying them out with actual forces.
- Exploring tactical interactions in simulation environments is necessary, given the increased importance of tactical cross-Service interoperation in current and future warfighting concepts. Both the Navy and the Marine Corps will need to participate in efforts to realize such simulation environments.
- Significant development of tools for building, validating, and verifying models; for generating scenarios; for populating databases; and for collecting and analyzing data is necessary. Furthermore such tools have to function, interoperate, and integrate into various environments and frameworks, as future experimentation campaigns are defined and executed.

These observations and the earlier discussions on platform availability and needed modeling capabilities lead to two findings.

Finding for Navy: The infrastructure and tools required for the experimentation campaigns of the future, including those for Sea Trial and joint experimentation, are inadequate. Primary shortfalls include the following: limited availability of ship platforms (compounded by the potential decommissioning of the USS *Coronado*) and airborne command, control, communications, computers, intelligence, surveillance, and reconnaissance (C4ISR) assets; lack of modeling and simulation capabilities that support a full set of experimentation campaign activities, explore tactical-level interactions, and reflect next-generation warfighting environments; and lack of tools for building, validating, and verifying models; for generating scenarios and populating databases; and for collecting and analyzing data.

Finding for Marine Corps: The infrastructure and tools required for the experimentation campaigns of the future are inadequate. Primary deficiencies include the following: lack of modeling and simulation (M&S) capabilities that support a full set of experimentation campaign activities, explore tactical-level interactions, and reflect next-generation warfighting environments; and lack of tools for building, validating, and verifying models; for generating scenarios and populating databases; and for collecting and analyzing data.

EXPERIMENTATION FOR BUILDING NAVAL FORCES FOR JOINT OPERATIONS

All recent U.S. military operations have been joint. Consequently the terms of reference for the study directed an examination of the role of experimentation in building future naval forces to operate in the joint environment. This section assesses experimentation relative to its influence on joint operations and with respect to preparedness for joint operations. Various experimentation venues are included—USJFCOM-sponsored experimentation activities, the Combatant Commands and cross-Service activities, as well as Naval Service activities.

Building Naval Forces to Support Joint Operations

Ultimately, the whole point of joint experimentation is better joint operations. Experimentation serves to help develop and refine concepts that lead to capabilities, which in turn are reflected in operations. Today, in support of joint operations, each Service brings its own core warfighting capabilities to the fight. Therefore, joint experimentation and Service experimentation both serve to build capabilities for joint operations. As indicated in Chapter 4, joint experimentation performs three primary functions in improving joint operations:

- Developing of DOTMLPF for the Services and the joint community,
- Improving the understanding of other Services' capabilities, and
- Contributing to the less tangible but important aspect of socialization.

To note progress first: The committee was consistently impressed by the major changes in thinking that have occurred among naval personnel within the past few years. With respect to understanding other Services' capabilities and to socialization, the Navy and Marine Corps officers now appear to have internalized the fact that they will almost always be fighting in joint contexts, often with much more integrated operations than would have been conceived of only a few years ago. This attitude is an important development in the evolution of U.S. military forces.

Experimentation is being planned more in terms of a joint context and with positive interest and much effort. The Navy and the Marine Corps realize that they may conduct key, theater-opening activities, with more massive Army and Air Force deployments following as quickly as possible. They also recognize the need to adapt to further shifts in traditional roles, as happened in Afghanistan when the Navy was present throughout the conflict and the Marines entered after the Army Special Operations Forces.

Although the joint experiments are in some respects a burden for the individual Services, they are also opportunities and, in any case, unavoidable realities. As a result, both the Navy and the Marine Corps are active participants and are increasingly building their own experiment programs around joint scenarios and working to align them with the schedule of joint exercises/experiments.[41] These efforts have led to increased collaboration.

There is also progress with respect to joint concept development,[42] although the migration of new concepts into actual operations is not so easy to trace, perhaps because it is too soon. However, the committee heard specific anecdotal examples of successes applied in Operation Enduring Freedom (OEF) that stemmed from activities connected with preparation for the USJFCOM MC 02 experiment; it heard of other successes in OEF stemming from the Hunter Warrior Advanced Warfighting Experiment. Lessons learned and equipment from experiments (specifically, the Marine Corps fast attack vehicles) were successfully used during combat in OEF.[43] There is also evidence that operational experience is helping to guide experimentation planning by USJFCOM.

Joint experimentation is still evolving and growing in importance. As of this writing, little development of DOTMLPF had occurred from USJFCOM-sponsored experimentation. Nonetheless, a few USJFCOM recommendations were close to having moved through the JROC approval process (e.g., the Standing Joint Force Headquarters and the Joint Enroute Mission Planning Rehearsal System—Near Term).

Collaboration in Concept Development

Although the Naval Services and the joint community are working toward joint concepts, the committee believes that interactions are not as collaborative as they need to be. As new warfighting concepts are developed, coordination among the Services and the joint community must ensure that individual concepts are compatible and supporting. This kind of coordination sets the stage for experimentation campaigns that develop capabilities and prepare forces for joint operations.

Concept development is focused for the Navy and the Marine Corps, respectively, in the Navy Warfare Development Command and the Marine Corps Combat Development Command. On the joint side, it occurs at the U.S. Joint Forces Command and, more recently, as part of the Joint Warfighting Capabilities Assessment (JWCA) process on the Joint Staff.

---

[41]See Table 3.1 in Chapter 3 and Tables 4.3 and 4.5 in Chapter 4. For instance, FBE-A included Air Force assets, as did FBE-D which added those of the Army. FBE-I was done in conjunction with Kernel Blitz, which had the participation of all four Services and the Marine Corps Capable Warrior experiment. The Navy had 10 experiment objectives incorporated in USJFCOM MC 02, including FBE-J, and the Marine Corps had 3, including Millennium Dragon '02.

[42]See Table 4.2 in Chapter 4.

[43]In Chapter 4, see the subsection entitled "The Role of Joint Experimentation in Preparing for Joint Operations."

While the NWDC and the MCCDC interact in concept development with USJFCOM and (along with other elements of the Navy) with the JWCA process, the committee believes that this interaction is not close enough. The committee's observations in this regard are as follows:

- The naval concepts being developed are generally couched in a joint context, but the extent of specific interaction with the joint community or other Services in developing and refining these concepts did not appear to be significant.
- The development of joint concepts by USJFCOM and the JWCA process does involve Naval Service representation, but even so, joint concepts seem to be created from scratch. They do not appear in a substantive way to build on the naval concepts that are being developed by the NWDC and the MCCDC. Likewise, there does not appear to be a detailed correspondence between the joint and naval-developed concepts.
- The joint community itself does not speak with a single voice, in that the various processes are developing different concepts. However, there is reason to anticipate that such differences will be resolved.[44]

Experimentation Led by the U.S. Joint Forces Command

USJFCOM has conducted a number of joint experiments of different types in which the Navy and Marine Corps have participated. These have included war games, constructive simulations, human-in-the-loop simulations, limited-objective experiments, and large-scale field experiments.[45] In addition, USJFCOM is planning a significant program of future experimentation across all types. This program will be characterized by two paths: (1) prototype refinement and validation, which have a nearer-term focus, and (2) concept development, which has a longer-term perspective.[46] The Navy and the Marine Corps are, of course, expected to be active participants in these future experiments.

To date, most discussion of USJFCOM-led experimentation has centered on the congressionally mandated field experiment Millennium Challenge '02, which was much larger than any prior joint experiment. The Navy and the Marine Corps each conducted its own major experiments—Fleet Battle Experiment-Juliet and Millennium Dragon '02, respectively—within the overall experiment. On the

[44]As this study was being completed, the Joint Staff was drafting the Joint Capstone Concept and the Joint Capabilities Integration and Development System document (CJCSI 3170.01C). Completion of the Joint Capstone Concept and its adoption by USJFCOM and realization of the processes specified in CJCSI 3170.01C could do away with the differences noted here.

[45]See Table 4.1 in Chapter 4.

[46]In Chapter 4, see the subsections entitled "Future Experimentation of the Joint Forces Command" and "Emerging Strategy of the Joint Forces Command for Experimentation Campaigns."

basis of briefings and other information provided by the Navy, Marine Corps, and USJFCOM, the committee makes the following general observations on Millennium Challenge '02:

- The value of such a large experiment relative to its cost[47] was questioned by a number of briefers and by many committee members. Some felt that funding could be put to better use in a series of smaller experiments both of the joint and the Service-specific types. Others noted, however, that such an event made the assets of other Services available, something that is difficult to achieve in the context of smaller experiments because of all the demands on operational assets. There was full agreement that it is extremely important to select the venue that matches the objectives desired. A larger-scale field experiment may be warranted to test integration, scalability, and a complex set of interactions, but the question remains—How large a scale is necessary?
- Such large, high-visibility experiments were characterized by many as demonstrations rather than as experiments, with little room for true exploration or failure. If that is the case, these large events may provide some opportunities to showcase unfunded but "ready-for-prime-time" equipment capabilities in hopes of garnering support and supplemental funding. Others noted, however, that a series of smaller events conducted over the 2 years leading up to Millennium Challenge '02 had allowed for greater exploration and assessment.
- Since the conduct of Millennium Challenge '02 in July and August 2002, the general sense in the community appears to be that further experiments of this size are unlikely. Unofficial statements from USJFCOM indicate that it will not conduct any more major joint experiments like Millennium Challenge '02 for the foreseeable future, but will focus its efforts instead on smaller, more frequent events.[48] The Navy and the Marine Corps should be active participants with USJFCOM in the planning now going on to define these future experiments, both for near-term prototype development and longer-term concept development.

Experimentation in the Combatant Commands

Joint experimentation also takes place in the Combatant Commands.[49] The issues involved in joint force operations are so important and large in scope that it is only natural that all Combatant Commands are involved in addressing them

---

[47]The committee repeatedly heard that the total cost of MC 02 was approximately $250 million, including the costs of the Services and of USJFCOM, but it has not been able to verify this number.

[48]Statement attributed to ADM Edmund Giambastiani, Commander of USJFCOM; see Anne Plummer, 2003, "Chief Says More Risks to Be Taken: USJFCOM Says No More Large-Scale Events Like Millennium Challenge," *Inside the Pentagon*, March 27, p. 3.

[49]In Chapter 4, see the section entitled "Experimentation in the Combatant Commands."

through experimentation. This has typically been done in combination with command post and field exercises.[50]

The Navy and Marine Corps and the Combatant Commands do cooperate in conducting joint experimentation within exercises conducted in the Combatant Commands—Kernel Blitz is one such example. However, there does not appear to be a coordinated program between the Services and Combatant Commands to conduct experimentation in a systematic way.

More collaboration of the Navy and Marine Corps and the Combatant Commands to systematically develop programs of joint experimentation appears highly desirable. Past experiences have been valuable, including many that applied to command, control, and communications; that pertained to developing and refining procedures or prototype systems; or that involved coalition nations. However, as indicated in Chapter 4, a number of activities would have to be carried out to achieve a greater degree of coordination in, and systematic expansion of, joint experimentation.

One possible mechanism for this coordination could be between the Combatant Command and the Service component commands assigned to that Combatant Command (e.g., either the appropriate Navy fleet command or the numbered fleet commands under that fleet command). The component commands could then interact as necessary with other elements of their respective Service (e.g., with the NWDC and the MCCDC).

Cross-Service Experimentation

Not all experimentation involving two or more Services need take place under the joint umbrella. Direct Service interaction can be very valuable. In the committee's view, experimentation involving two or more Services is best suited for operations at the tactical level, since higher-level (component and Joint Task Force) operations are primarily joint. Useful concepts developed in this manner can be fed into the joint arena. There is a need to investigate joint interactions at a tactical level, given the growing intensity in recent operations, such as that in Afghanistan. While the Navy and Marine Corps have devoted increased attention to cross-Service experimentation, more is needed.

Direct Service interaction can occur in two general ways—between Service centers involved in concept development and between deployed forces engaged in exercises and experiments. The Navy Warfare Development Command and the Marine Corps Combat Development Command have regular interactions with one another. In addition, the Navy Network Warfare Command has noted that its proximity to corresponding Air Force organizations (e.g., Air Combat Command

[50]Table 4.5 in Chapter 4 provides several examples.

and the Air Force C2ISR Center) offers the opportunity for collaboration. While other, substantive interactions between Service centers may exist, none was apparent to the committee in its investigations. Increased interaction could be very valuable—to provide cross-fertilization in concept development and to begin building joint concepts from the ground up.

The Navy indicated in briefings to the committee that other Services (e.g., the Air Force) are involved in its fleet exercises and that it would like to increase such involvement. Recent Marine Corps experiments, with their highly tactical focus, do not appear to have a significant joint or cross-Service perspective. However, future opportunities would appear to exist—for example, coordinated operations with Army Special Operations Forces and air support from Navy and Air Force aircraft.

One recurrent difficulty with cross-Service experimentation is the need for the simultaneous availability of assets from two Services. Mutual scheduling is difficult. Ideally, the elements of both Services should be on exercises or engaged in other more formal experiments at the same time. This, however, requires cross-Service coordination of force deployment schedules. Such coordination would have to be carried out at the top levels of the Services—possibly with USJFCOM acting as an intermediary. This would be necessary because the deployment schedules have Service-wide impacts relating to such factors as accomplishing operational missions, training, personnel rotation, and platform overhaul and upgrade. Such a coordination mechanism is not in place today.

Balancing Service and Joint Experimentation

Since naval operations will become more joint in coming years, linkages between naval and USJFCOM sponsored experiments, and indeed the full range of joint experimental campaigns, should be carefully planned. This spectrum of activities runs from the earliest concept development, through analysis, war games, and simulations, and leads ultimately to LOEs and large field experiments. An orderly progression of these activities will greatly aid in the efficient use of Service and joint resources.

In all of these phases, joint experimentation coexists with Service-specific experimentation because proficient Service core capabilities are a prerequisite for joint warfighting. Alternately, joint experimentation and the attendant joint concept development may require that the Services develop new capabilities. Service experimentation programs should progress in an orderly fashion so that they can feed into and contribute to joint experimentation while also deriving benefits from the joint activities—all the better for both Service and joint capabilities.

USJFCOM has laid out a detailed and intense joint experimentation campaign. The Services support and participate in this effort while in turn performing their own Service-unique experimentation activities. These multiple responsibilities stretch the already resource-limited Service-specific experimentation efforts. It has already been noted that the MC 02 event significantly stressed the staff resources of the Marine Corps. If, in the worst case, USJFCOM's activities interfere with or prevent the normal progression of Service experimentation, they can actually be counterproductive to developing improved joint capabilities. Thus, a balance must be maintained between joint and Service-unique experimentation, with events and objectives synchronized within the respective experimentation campaigns. To date there is no mechanism in place for achieving this balance.

Summary with Findings

In assessing the role of experimentation in building forces for joint operations, the committee noted progress both for the Navy and the Marine Corps. The following limitations and opportunities also exist in the current situation:

- There is a need for expanded and more synergistic collaboration in joint concept development.
- There are opportunities for expanding joint experimentation through the Combatant Commands, but a coordination mechanism is needed to develop programs of experimentation systematically.
- There are opportunities and a need for more cross-Service experimentation, including investigating joint interactions at the tactical level (particularly important for the Marine Corps), given recent operations such as that in Afghanistan. However, a mechanism is required to schedule the simultaneous availability of assets for the Services involved.
- There is a need to balance and synchronize joint and Service-unique experimentation, given the demands placed on the resources available.

These observations lead to the following finding.

Finding for the Naval Services: The naval and joint experimentation programs are not yet adequately aligned and synchronized, nor is there sufficient correlation between them. More synergistic collaboration in joint concept development is needed. Service-unique experimentation has been and could continue to be affected by large-scale joint experimentation, but no formal mechanism exists for striking a proper balance between joint and naval experimentation.

6

Recommendations for Improving the Overall Effectiveness of Naval Experimentation

This study identifies many fruitful areas of recent naval experimentation. They include, for example, the evolution of the fleet's network-centric operations, the Marine Corps's Urban Warfare doctrine, and the developing concept of the Littoral Combat Ship (LCS). Past and recent efforts are summarized in Chapters 3 and 5. Nonetheless, there are several issues with respect to Navy and Marine Corps experimentation that require resolution if experimentation is to be an enabler for naval force development. These issues, identified as findings, are discussed in detail in Chapter 5.

What follows in this chapter are the recommendations of the committee in response to these findings. For the convenience of the reader, a synopsis of the issue leading to a finding and the finding itself precede the associated recommendation. The order of the recommendations follows the order of findings as presented in Chapter 5—with one exception. The committee decided that the recommendation regarding senior Navy oversight is foremost. In the aggregate, the recommendations consist primarily of new or enhanced strategies, mechanisms, and processes and are pertinent to the ongoing and future experimentation programs of the Naval Services in this period of rapid change in their approach to their mission.

ESTABLISH SENIOR NAVY OVERSIGHT AND ANNUAL REVIEW OF EXPERIMENTATION EFFORTS

Issue

The Chief of Naval Operations (CNO) is using experimentation as an enabler for his future vision. He has expectations that experimentation will contribute to building future forces, provide the means for the development of advanced concepts, and facilitate the transfer of capabilities to the field. What the committee sees is a modest program of experimentation managed by an organization—the Navy Warfare Development Command (NWDC)—with insufficient influence over the Navy experimentation program[1] and its numerous participants, including the funding and assets required, and with insufficient leverage to move the results of experimentation either into acquisition or into programs of record through the requirements process. This situation exists in a very large institution in which many competing priorities exist and many bureaucracies operate.

Given the past lack of control over required funding, assets, and use of results, the committee debated the need for direct alignment of the NWDC under the CNO in order to engender the necessary response to experimentation within the Navy as well as to boost the effectiveness of experimentation. The CNO recently realigned the NWDC under the Commander, Fleet Forces Command (CFFC), and it is difficult to ascertain if that step will improve the current situation. The committee was concerned that the realignment would give rise to another issue—that of undue focus on near-term and mid-term objectives, resulting in an unbalanced program that would not meet the CNO's long-term objectives. Given the current situation, the committee concluded that the sustained attention of the CNO and Vice Chief of Naval Operations (VCNO) is vital for the overall coordination, direction, prioritization, and execution of the naval experimentation program.

Finding for Navy: Continuing Chief of Naval Operations and Vice Chief of Naval Operations sponsorship, leadership, and attention are vital for the overall coordination, direction, prioritization, and execution of the naval experimentation program.

Recommendation 1: To ensure that experimentation is a key enabler of his long-term vision, the CNO should establish and, together with the VCNO, participate in an annual review of the experimentation program with the senior leadership of the Navy. This process should make visible the program content; the balance of near-term, mid-term, and long-term objectives; the progress that has occurred to

[1]There is no one formal Navy experimentation program; instead a number of organizations are engaged in experimentation activities that collectively embody a Navy program. In Chapter 3, see the subsection entitled "Organizational Roles and Major Participants in Navy Experimentation."

date; the results that have been achieved; the use that has been made of the results relative to doctrine, organization, training, materiel, leadership, personnel, and facilities (DOTMLPF), including the transition to requirements, acquisitions, programs of record, and/or capabilities in the fleet; and guidance for the future.

STRENGTHEN TRANSITION PROCESSES

Issue

The mechanisms and processes for transitioning the results of experimentation, including transition planning, need strengthening. For the Navy, despite the transitioning of some concepts, doctrine, and TTPs to the field, there is little evidence that equipment and capabilities resulting from experimentation have transitioned or are directly linked to major acquisitions and programs of record. Although the results of experimentation have had some influence on acquisition designs (e.g., for the Littoral Combat Ship), these instances appear to have evolved through personal connections rather than through institutional mechanisms. In general, institutional mechanisms are preferred because they typically endure beyond personal connections.[2]

In the Marine Corps, more structured transition planning is evident, as are some successes in moving concepts, doctrine, TTPs, and minor items of equipment into the field. Nonetheless, there was no evidence that major items of equipment from experiments had resulted in major acquisitions or displaced programs of record. Although not all experimental capabilities warrant transition in the context of cost, risk, and military value, the debate should be allowed. The committee also noted that spiral development—a potential enabler for transitioning capabilities more rapidly to the fleet—has not been explored systematically or incorporated as a fundamental method of experimentation.

Finding for Navy: The mechanisms and processes for transitioning the results of experimentation directly to the fleet or to an acquisition program of record are inadequate, and they curtail the effectiveness of experimentation in building future naval forces.

Finding for Marine Corps: The Marine Corps has been successful in transitioning nonmaterial elements of doctrine, organization, training, materiel, leadership, personnel, and facilities (DOTMLPF) and minor equipment. However, it has not been successful in transitioning to combat forces major warfighting capabilities identified during experimentation.

---

[2]However, there are times when a personal connection is the initial enabler and the CNO can often assign or rotate key personnel to maximize benefits.

Recommendation 2: To strengthen the transition of experimentation results to the requirements and acquisition processes, the Secretary of the Navy, the Chief of Naval Operations, and the Commandant of the Marine Corps should institute specific procedures to facilitate and accelerate the transition of capabilities identified through experimentation to the fleet.

Specifically, for the Navy:

- The Commander, Fleet Forces Command; Deputy Chief of Naval Operations for Warfare Requirements and Programs (N6/N7); Deputy Chief of Naval Operations for Resources, Requirements, and Assessments (N8); and Assistant Secretary of the Navy for Research, Development, and Acquisition should collectively formalize a planning process for the transition of the operational and system capabilities emerging from experiments to the fleet. The process should include framing transition issues and identifying potential funding gaps.
- The N6/N7 and the N8 should develop a process which ensures that the successful results of experiments are adequately evaluated and competed with the programs of record in the context of cost, risk, and military value.
- The Navy operational and acquisition communities should explore means to accelerate transition of the results of experimentation to the fleet more aggressively. These means should include the expanded use of other transaction authority and spiral development.[3]
- The Navy test community should explore new roles for the Operational Test and Evaluation Force, including its early participation in the experimentation program, with its advisory assessments provided directly to experiment managers.

Specifically, for the Marine Corps:

- The Marine Corps Combat Development Command, in conjunction with the Marine Corps Systems Command, should expand early transition planning in order to include the framing of transition issues and the identification of potential funding gaps.

---

[3]The use of spiral development to accelerate deployment of capabilities to the fleet has not been systematic to date, although spiral development is a component of "Sea Power 21." See Chapter 5 for additional details. Both the Air Force and Army have enjoyed some notable successes by incorporating it into their respective experimentation programs. See Chapter 3 for additional details. Given the Navy's emphasis on network-centric operations and NETWARCOM's emerging role, the committee believes that spiral development should be explored through experimentation to accelerate network-centric capabilities into operations. There are also naval infrastructures that may support such an exploration—namely, the Navy's Distributed Engineering Plant and the Marine Corps's Tactical Systems Support Activity.

- The Marine Requirements Oversight Council should establish a process which ensures that the successful results of experiments are adequately evaluated and competed with the programs of record in the context of cost, risk, and military value.
- The Marine Corps operational and acquisition communities should explore means to accelerate transition of the results of experimentation to the fleet more aggressively. These means should include expanded use of other transaction authority and spiral development.
- The Marine Corps test community should explore new roles for the Marine Corps Operational Test and Evaluation Activity emphasizing its early participation in the experimentation program, with its advisory assessments provided directly to experiment managers.

Recommendation 2 strengthens the connections between experimentation and the requirements and acquisition processes in three ways: (1) through early formal transition planning to frame transition issues and identify potential funding (for the Marine Corps it expands the current planning process with early involvement of the Marine Corps Systems Command in issues and funding); (2) through the establishment of a process which ensures that the successful results of experiments are adequately evaluated and competed with the programs of record in the context of cost, risk, and military value; and (3) through the exploration of means to accelerate the results of experimentation to the field more aggressively. Included in the third way of strengthening these connections are expanded use of other transaction authority, use of spiral development, and participation of the test community in experimentation, but in nontraditional roles.

The committee used lessons learned about experimentation from the Army and the Air Force. For instance, the committee noted a recent and promising change in funding processes by the Air Force,[4] established to ensure that a source of funding is available for experimental capabilities. As part of early Air Force transition planning, promising initiatives are vetted Air Force-wide to focus and pare them and to establish exit criteria for transition. When approved, an initiative is associated with a program element in the Program Objectives Memorandum (POM) that would subsume and fund the initiative if it proved successful. In addition, the Air Force maintains a modest source of funds to bridge experimentation initiatives until the identified POM funding stream becomes available. This two-part strategy overcomes the lack of programmed resources for implementing initiatives when the results of experimentation prove successful.

Another strategy employed by both the Army and the Air Force is that of participation in experimentation by the Services' operational test communities,

[4]In the discussion of Air Force Experimentation in Chapter 3, see the subsection entitled "From Experiments to the Field."

but in nontraditional roles. As an example, the Air Force Operational Test and Evaluation Center (AFOTEC) is involved in development evaluation and now participates at the beginning of experimentation. It provides an assessment of an experimental capability directly to a program manager/developer, but this type of assessment is not the usual operational test for acquisition of a production capability. This earlier development assessment is non-threatening; it points out issues that need resolution through experimentation and identifies key problems such as critical safety issues, while familiarizing the test community with how operators use a promising new capability (facilitating later testing). If the experimental capability is successful, at the appropriate time the AFOTEC evaluates the initiative again and conducts the usual testing and evaluation required to support the formal acquisition process.

The Naval Services should also consider expanded use of other transaction authority (OTA)[5] for experimentation. It is an acquisition strategy that is applicable to prototype projects, provides streamlined procedures for faster awards, and enables innovative business arrangements that can transition later into the Major Defense Acquisition Program processes.

Reaffirmation of Spiral Development

A component of Recommendation 2 reaffirms the promise of spiral development that was first articulated in the recent Naval Studies Board report on network-centric operations.[6] The use of spiral development by the Naval Services has not been systematic to date. Certainly, given the issues in transition, this is an appropriate reiteration of a prior recommendation.

Both the Air Force and the Army have enjoyed some notable successes by incorporating spiral development into their respective experimentation programs. (In Chapter 3, see the subsections entitled "U.S. Army" and "U.S. Air Force" for additional details.) As an alternative to a lengthy and detailed requirements-based process, both Services established integration environments with operators and developers collaborating on experimental capabilities with the intent of delivering a core capability quickly and then fielding additional capabilities subsequently and incrementally. The results are mixed, but there are some notable successes,[7]

---

[5]Jacques S. Gansler, Under Secretary of Defense for Acquisition, Technology, and Logistics. 2001. "Other Transactions (OT) Guide for Prototype Projects," The Pentagon, Washington, D.C., January. Available online at <http://www.afmc.wpafb.af.mil/HQ-AFMC/PK/pkt/OTGuideAug2002.doc>. Accessed October 10, 2003.

[6]Naval Studies Board, National Research Council. 2000. *Network-Centric Naval Forces: A Transition Strategy for Enhancing Operational Capabilities,* National Academy Press, Washington, D.C.

[7]For details, in Chapter 3, see the subsections entitled "A Past Example of Army Experimentation," and "Examples of Air Force Experimentation."

particularly for capabilities heavily dependent on information technologies, such as command and control, communications, and networks.

Given the leadership responsibility of the Navy Network Warfare Command in network-centric operations and its emerging role in experimentation efforts related to this mission, the committee believes that spiral development should be intrinsically coupled with experimentation to accelerate network-centric capabilities into operations. There are also naval infrastructure capabilities that could support such exploration—namely, the Navy's Distributed Engineering Plant and the Marine Corps Tactical Systems Support Activity at Camp Pendleton, California.

ENHANCE THE NAVAL EXPERIMENTATION PROGRAMS

Issue

Certain important areas are not yet adequately explored in the naval experimentation programs, although some of these areas are gaining definition.[8] For the Navy, these omissions are due in part to its approach to experimentation, which in the past has not been founded on sufficiently robust experimentation campaigns but on an over-reliance on many individual events such as FBEs. Such singular events cannot provide the depth of knowledge required to explore potential concepts and capabilities sufficiently. Not only has the breadth of these programs been limited, but the number of concepts explored has been small, the concepts have not covered a sufficiently broad range, and they have not been systematically chosen and developed. As a result, the experimentation programs have lacked the cohesion and comprehensiveness needed to address the challenges of Sea Power 21 or to deal conclusively with questions about capabilities that will be delivered by the programs of record.

Areas that need further investigation include over-the-horizon, time-critical strike; use of extended-range guided munitions for long-distance, high-volume, rapid fire support; expanded applications of network-centric capability to deployable undersea sensor arrays; mine/countermine warfare; and the use of unmanned aerial vehicles to locate and identify targets. The Navy's experimental work to date has brought out overarching issues such as the need to achieve a satisfactory common operating picture; deconfliction; and bandwidth size and

[8]In January 2003, the CNO requested that the CFFC—as part of its lead role for Sea Trial in support of Sea Power 21—"[d]raft and implement a comprehensive roadmap (by May 2003) that integrates studies, wargames, experimentation, and exercises with evaluation metrics and an execution timeline." See Chief of Naval Operations, 2003, *CNO Guidance for 2003,* Department of the Navy, Washington, D.C., January 3. Available online at <http://www.chinfo.navy.mil/navpalib/cno/clark-guidance2003.html>. Accessed November 9, 2003.

management. Some important areas not yet explored include Vertical Launch System reloading at sea, assault breaching of mine or obstacle fields near and on the beach, and continued decisive operations under impaired network conditions and unfavorable environmental conditions.

In response to the CNO's guidance, the CFFC, through the NWDC, recently drafted the Sea Trial experimentation campaign plan.[9] The committee believes that this is a step in the right direction, although the impact of the plan on the Navy is as yet unclear.

For the Marine Corps, there has been a shift in recent years from a balanced program of experimentation campaigns to a program of experimentation based on near- and mid-term objectives. While these immediate challenges are important, there remains a need for continuing investment in long-term experimentation. Examples of areas to examine include sea basing, for which a program of experimentation needs to be designed, funded, and executed with the objective of realizing new capabilities, doctrine, and TTPs; operations in brown-water littorals to negate potential threats; and unconventional warfare, for which current procedures will have to be adapted for use from a sea base in brown-water operations.

Finding for Navy: As yet, no cohesive experimentation program exists that will move the Navy's forces to "future" concepts, processes, doctrine, and capabilities.

Finding for Marine Corps: The Marine Corps has moved from a balanced program of experimentation campaigns based on near-term, mid-term, and long-term objectives to one of experiments focused on near- and mid-term objectives.

Recommendation 3: To address strategic, long-term objectives of Sea Power 21 and Expeditionary Maneuver Warfare, the Department of the Navy should expand its programs for experimentation.

Specifically, for the Navy:

- The Commander, Fleet Forces Command, with the support of the Navy Warfare Development Command, should (1) create and maintain updated experimentation campaigns that address transformation objectives while identifying actionable steps and the organizations responsible for them; (2) ensure a balance in experimentation efforts directed at near-, mid-, and long-term objectives; (3) conduct experimentation sufficient to ensure that the highest-priority operational concepts are explored adequately for incorporation into the fleet and its operations; (4) establish adequate mechanisms for continued improvements and

[9]Commander, Fleet Forces Command. 2003. *Sea Trial—Concept Development and Experimentation Campaign Plan (U),* Working Paper (draft), Norfolk, Va., May (Classified).

modifications to the experimentation program; and (5) maximize the effectiveness of joint experimentation in accordance with Recommendation 7 (below).

• The Navy Network Warfare Command and its supporting organizations should play a lead role in coordinating the information network aspects of experimentation and in enabling the realization of network-centric capabilities for the fleet through related concept development or exploration and spiral development processes.

Specifically, for the Marine Corps:

• In collaboration with the Navy Warfare Development Command, the Marine Corps Combat Development Command should augment its experimentation program by developing experimentation campaigns to address its strategic, long-term objectives, such as sea basing, conventional and unconventional expeditionary warfare, and, jointly with the Navy, assured access. Furthermore, its overall experimentation campaign should encompass all levels of force structure and activity necessary to meet the range of potential threats and future operational demands.

The committee believes that the expanded use of experimentation campaigns, balanced between near-term, mid-term, and long-term objectives, is a needed shift in strategy.

ENHANCE NAVY EXPERIMENTATION PROCESSES

Issue

Some omissions in the Navy's program of experimentation, noted above, do result from inadequate methods in building experimentation campaigns and from the underutilization of some experimentation venues. To date, preparing for the event of an FBE has largely (but not exclusively) been a focus of activities. The NWDC needs both to shift focus and to augment processes.

The NWDC requires enhanced processes to select concepts for exploration; to integrate more overarching studies and analyses throughout the events of its campaigns; to build, mature, and evaluate concepts, including multiple and competing concepts; and to maximize and apply a full range of experimentation venues (such as games, modeling and simulation, and limited-objective experiments) in a more systematic manner.

Finding for Navy: There are significant deficiencies in the end-to-end processes of naval experimentation. Shortfalls include the following:

- Insufficient use of structured experimentation campaigns;
- Inadequate breadth in exploration of concepts;
- Insufficient studies and analysis, including their use to determine the best experimentation venue (games; simulations; small, focused experiments; and so on);
- Inadequate use of spiral processes to build knowledge iteratively; and
- Inadequate planning and evaluation.

Recommendation 4: To improve the effectiveness of its experimentation efforts, the Navy Warfare Development Command should augment its end-to-end experimentation processes by making the following key changes:

- Expand the emphasis on experimentation campaigns that use a full spectrum of experimentation activities, with analysis integrated throughout the campaigns as well as applied to determine which venues are most appropriate.
- Conduct significantly greater amounts of systematic and innovative analysis earlier and throughout the experimentation process in order to select, develop, and broaden understanding of the operational concepts to be explored, including a range of multiple and competing concepts.
- Broaden the incremental, sequential approach by using spiral methodology. Apply the sequential approach to war games, to modeling and simulation, and to small-scale, more narrowly focused experiments. Build larger-scale experiments on the basis of the results of such a sequential approach.
- Establish a standing, high-level, independent technical advisory board composed of experts in methods of innovation and experimentation and reporting to the Commander of the Navy Warfare Development Command, as a means to foster more robust experimental processes, maintain their quality, and make recommendations for improvements.

The committee believes that building good experimentation campaigns is a key to a more comprehensive naval experimentation program. However, campaigns must be well structured so that they explore concepts in depth, investigate multiple and competing concepts, and "write the book" on knowledge about the concepts and capabilities under investigation. Enhancing current NWDC processes in a few areas will build this more robust understanding. These enhancements include expanded use of analysis throughout campaigns, including the evaluation plan and selection of the appropriate experimentation venue, and the use of spiral processes that build upon each venue to refine or discard preconceived assumptions. The committee notes progress in this direction in the recent Sea Trial Experimentation Plan prepared by the NWDC.

These recommended changes shift emphasis onto experimentation events other than FBEs. The committee also applauds the recent Sea Trial Experimentation Campaign Plan for its inclusion of a greater number of small experimentation venues. While large field experiments are important and have value, they should be used only when they are the appropriate event—for instance, when issues of scalability or integration have to be explored, or when warfighters need to interact with potential new capabilities. The changes will achieve better balance across all venues and will build knowledge, such as that gained from a large-scale examination of options or excursions, that cannot be learned in the large field experiments.

SUSTAIN AND USE NAVY EXPERIMENTATION RESOURCES MORE EFFECTIVELY

Issue

The committee questioned whether there are sufficient resources available to the NWDC for experimentation, but it found this difficult to answer for two reasons. One is the past emphasis on large field events, compounded more recently by joint experimentation. Large field events are costly and require many months of effort.

The second reason is that the NWDC does not have line-item funding to cover the full costs of its own experimentation program but relies on additional funding from many organizations, such as the Office of Naval Research. These sponsors may reallocate funds in accordance with their own higher-priority efforts. Such perturbations have severe impacts on the planning, preparation, conduct, and evaluation of experiments. The changes in already-programmed funding are frequent and severe enough to make the need for additional funding unclear. This situation leads to a poor return on investment and results in delays and/or some loss of effort. Consequently, the Navy is not making the most effective use of the experimentation resources already programmed.

Finding for Navy: The Navy has not made effective use of resources in its experimentation program. Specifically:

- The Navy has overemphasized large field experiments, when in some cases smaller, better-focused venues would have served equally well and would have avoided the compromise of objectives often associated with experiments conducted in combination with training exercises and maintenance schedules.
- The Navy has inadequately preserved and protected resources intended for experimentation.

Recommendation 5: The Navy should use the resources already programmed for experimentation more effectively, while also resolving resource contention surrounding experimentation. Specifically:

- The Chief of Naval Operations, with input from the Vice Chief of Naval Operations; the Commander, Fleet Forces Command (CFFC); the Deputy Chief of Naval Operations for Warfare Requirements and Programs (N6/N7); and the Deputy Chief of Naval Operations for Resources, Requirements, and Assessments (N8), should determine, and then plan, program, and preserve sufficient funding for the experimentation program within the Future Year Defense Program as a matter of priority. Funding for the Navy's experimentation program should be a separate line item that is not commingled with funds provided by other Navy sponsors, such as the Office of Naval Research.
- The CFFC should ensure that sufficient priority is given to experimentation needs when fleet training exercises and maintenance events override or threaten to compromise experiments. The CFFC should oversee and adjudicate conflicts that arise during all stages of fleet experiments.
- In collaboration with the CFFC and the Navy Warfare Development Command (NWDC), the Chief of Naval Education and Training, the Naval Postgraduate School at Monterey, California, and the Naval War College at Newport, Rhode Island, should develop a tailored course on experimentation in order to train and educate sailors, Marines, and civilians, and to instantiate best practices in experimentation.

In summary, by shifting to a more balanced use of all experimentation venues, inappropriate reliance on the more costly field events is decreased and a more effective return on investment is achieved. FBEs will continue to be an important component of experimentation, but they will be used where appropriate. Their continuation means that the attendant contentions over resources, such as sharing assets with training exercises, needs to be addressed. The CFFC's command of fleet assets should provide resolution and leverage in this matter.

The answer to the perturbations in available funding for experimentation is an important matter. Whatever the level of funding deemed appropriate by the Navy, the NWDC should be able to count on that and plan accordingly. While sponsors outside the Navy may contribute other assets and funding, the Navy funding should be sustained at the agreed-to level.

Lastly, to maximize the effectiveness of people engaged in experimentation, the Navy should establish and administer a course on experimentation. While such a course does not appear to be currently available, it would build expertise, provide insight into lessons learned, and instantiate best practices in experimentation.

ENHANCE INFRASTRUCTURE AND TOOLS FOR NAVAL EXPERIMENTATION

Issue

The Naval Services are in need of improvements in the infrastructure and tools supporting experimentation. Some critical platforms for experimentation are unavailable, including ship platforms—a shortfall compounded by the potential decommissioning of the USS *Coronado*—and airborne command, control, communications, computers, intelligence, surveillance, and reconnaissance (C4ISR) assets, some of which are currently under the ownership or control of the other Services.

In addition, the Naval Services will require improvements in modeling and simulation (M&S) tools to explore the full range of experimentation venues, including tools for generating scenarios, populating databases, and collecting and analyzing data. Simulations are required to explore more tactical-level interactions. At the conclusion of the Millennium Challenge '02 exercise, the U.S. Joint Forces Command (USJFCOM) noted that the Defense Department's existing M&S capabilities are insufficient to represent future operational concepts. In response, USJFCOM is proceeding with the expansion of its own Continuous Experimentation Environment. As a result, the Navy and Marine Corps will have an opportunity to leverage the USJFCOM investments while also ensuring compatibility across the respective Service and joint environments. Industry partnerships can be explored also, as appropriate.

Finding for Navy: The infrastructure and tools required for the experimentation campaigns of the future, including those for Sea Trial and joint experimentation, are inadequate. Primary shortfalls include the following: limited availability of ship platforms (compounded by the potential decommissioning of the USS *Coronado*) and airborne command, control, communications, computers, intelligence, surveillance, and reconnaissance (C4ISR) assets; lack of modeling and simulation capabilities that support a full set of experimentation campaign activities, explore tactical-level interactions, and reflect next-generation warfighting environments; and lack of tools for building, validating, and verifying models; for generating scenarios and populating databases; and for collecting and analyzing data.

Finding for Marine Corps: The infrastructure and tools required for the experimentation campaigns of the future are inadequate. Primary deficiencies include the following: lack of modeling and simulation (M&S) capabilities that support a full set of experimentation campaign activities, explore tactical-level interactions, and reflect next-generation warfighting environments; and lack of tools for building, validating, and verifying models; for generating scenarios and populating databases; and for collecting and analyzing data.

Recommendation 6: To investigate a full range of experimentation venues, the Department of the Navy should enhance its infrastructure and tools for naval experimentation.

Specifically, for the Department of the Navy:

- Given the importance of having a command ship platform with an expert and experienced cadre of experimentalists for work in network-centric operations, the Department of the Navy should ensure that the USS *Coronado* (or a comparable platform) remains available for experimentation as it makes changes in the fleet of ships in commission or in active reserve duty.
- The Department of the Navy should augment its modeling and simulation tools and infrastructure to support a full spectrum of naval and joint experimentation campaign activities (e.g., concept development, war games, and limited-objective experiments) to support tactical-level interactions and to reflect next-generation warfighting environments. It should also supplement its tools for building, validating, and verifying models; for generating scenarios and populating databases; and for collecting and analyzing data, while ensuring that tools can function and integrate within the various frameworks and environments as future experimentation campaigns are defined and executed. In addressing enhancement of its infrastructure and tools for naval experimentation, the Department of the Navy should leverage the capabilities of other organizations, such as the U.S. Joint Forces Command.

Limited resources are available for experimentation, and the augmentation of infrastructure and tools can exact a heavy funding toll. The committee believes, however, that enhancements are necessary. The first bullet item in Recommendation 6 addresses the need for a command ship platform that also brings expertise for experimentation. The USS *Coronado* has provided excellent support in the past, but the Navy should best determine how to fill the void that would be left by the ship's decommissioning—for example, relegate it to some other status or designate some alternate platform for experimentation.

The committee anticipates that the realignment of the NWDC under the CFFC should provide more "muscle" for acquiring the assets required for experimentation. CFFC is in a stronger position than is the NWDC to adjudicate when there is competition for assets within the Navy and to work with the other Services for access to platforms not owned by the Naval Services, such as key aerial C4ISR vehicles.

The second bullet item in Recommendation 6 primarily addresses tools. Augmenting tools for experimentation, particularly M&S enhancements for future campaigns, is required for both joint and naval experimentation. Two strategies can and should be used to mitigate the funding impacts—the strategy of leveraging USJFCOM's efforts to evolve its own Continuous Experimentation Environment[10] (thereby ensuring compatibility), and that of prioritizing the enhancements in accordance with their "need dates" as experimentation campaigns are developed for the future.

BALANCE NAVAL AND JOINT EXPERIMENTATION

Issue

As joint operations continue to leverage naval capabilities, planning is needed to link naval and joint experimentation across a spectrum of activities ranging from the earliest concept development through analyses, war games, and simulations, leading ultimately to limited-objective experiments and larger fleet experiments. Given this requirement, the committee views the current state in naval experimentation as having both limitations and opportunities.

Joint concept development requires more synergistic collaboration with the Naval Services than is now taking place. Concept development for the Navy and the Marine Corps is focused at the NWDC and the MCCDC, respectively, and on the joint side at the USJFCOM. Although these organizations are interacting, the committee believes that the interaction is not as close or extensive as it needs to be. While the development of joint concepts does involve the Naval Services, it does not appear to build on the Services' concepts in a substantive way; nor is there a good, detailed crosswalk between the joint concepts and those developed by the Naval Services.

In the past the Navy and Marine Corps have participated with USJFCOM in a number of joint experimentation events ranging from war games to field experiments. USJFCOM is currently planning a significant program of future experimentation activities. The Naval Services are expected to be active participants and should play a substantive role in defining them. There are also opportunities to expand joint experimentation through the Combatant Commands and through cross-Service activities, the latter particularly necessary to investigate joint interactions at the tactical level (though recent Marine Corps experiments with their tactical focus do not appear to have significant cross-Service or joint perspective). However, a mechanism is needed to balance and synchronize joint and naval experimentation, given the demand placed on the resources available.

---

[10]Discussed in detail in Chapter 4, in the subsection entitled "Synopsis of Results of Joint Forces Command Experimentation to Date."

Finding for the Naval Services: The naval and joint experimentation programs are not yet adequately aligned and synchronized, nor is there sufficient correlation between them. More synergistic collaboration in joint concept development is needed. Service-unique experimentation has been and could continue to be affected by large-scale joint experimentation, but no formal mechanism exists for striking a proper balance between joint and naval experimentation.

Recommendation 7: To ensure better preparation for future joint operations and to maximize the effectiveness of its participation in joint experimentation, the Department of the Navy should establish a set of principles and guidelines to balance experimentation requirements for Service-unique and joint experimentation, and it should then align and synchronize its participation accordingly. Specifically:

- The Chief of Naval Operations (CNO) and the Commandant of the Marine Corps (CMC) should actively support joint experimentation on the basis of a clear understanding of the priorities of the various joint concept development and experimentation activities. In addition, they should advocate top-level interest in operational concepts driven by naval force capabilities as well as concepts suitable in other operational environments.
- The Commander, Fleet Forces Command (CFFC) and the Marine Corps Combat Development Command (MCCDC) should conduct enough naval experimentation campaigns to ensure that the highest-priority naval operational concepts are adequately explored.
- The CFFC and the MCCDC should design all naval experiments with full recognition that the Navy and the Marine Corps will most likely be operating in a joint context; to the maximum degree feasible, the Naval Services should partner with the other Services in experimenting with relevant assets.
- The NWDC and the Marine Corps Warfighting Laboratory should achieve adequate cross-fertilization of joint and naval-specific operational concepts through substantive interaction of the respective concept development communities.
- The CFFC and the MCCDC should support and participate in joint experiments to explore the interaction and mutual support of the future operational concepts of each of the Services. These efforts should include staff interactions at the operational level and the "removal of seams" between components at the tactical level.
- The CFFC and the MCCDC should work with USJFCOM to identify key challenges (e.g., cruise missile defense, joint intelligence, surveillance, and reconnaissance) on which they could welcome joint and/or other Service contributions.
- The CFFC, N7, N8, and the MCCDC should examine the tradeoffs in the benefits of large, resource-intensive and less well controlled experiments against the opportunities lost by not conducting a greater number of smaller, more focused joint and/or naval experimentation activities.

- The CFFC and the MCCDC should systematically develop programs of joint experimentation with the Combatant Commands and should establish a coordination mechanism to facilitate the development of such programs.
- The CFFC and the MCCDC should increase cross-Service experimentation, particularly at the tactical level, and should establish a scheduling mechanism to facilitate this experimentation.

The committee's Recommendation 7 expounds on opportunities for joint experimentation, not only through USJFCOM, but also with the Combatant Commands and through cross-Service efforts. In the past both the Navy and the Marine Corps have participated in conducting experiments with the Combatant Commands, usually in conjunction with exercises such as Kernel Blitz. Past experiences have been valuable, particularly with respect to advances in command and control, communications, developing and refining procedures and working with prototype systems, and the involvement of coalition nations. Nonetheless, as noted, there is a need for a coordination mechanism that will operate between the Naval Services and the Combatant Commands to identify and build programs of mutual interest. One alternative is to implement coordination between the Combatant Command and the Service component commands assigned to it.[11] The component commands could in turn interact with appropriate elements of their respective Services (e.g., with the NWDC and the MCCDC).

More cross-Service experimentation addresses a pressing need to investigate joint interactions at the tactical level, given the growing intensity of such interactions in recent operations. This is particularly important for the Marine Corps. Such future experimentation opportunities would appear to exist—for example, in coordinated operations with Army Special Operations Forces and air support from Navy and Air Force aircraft. Expanding cross-Service experimentation, however, requires a mechanism for coordinating force deployment schedules. It is conceivable that such coordination could be carried out at the top levels of the Services, with USJFCOM serving as an intermediary.

The committee's recommendation above provides a set of principles and guidelines to use in striking a balance between Service-unique and joint experimentation. While these principles are suggestions by the committee, the Navy and the Marine Corps can develop and evolve their own set, and then use it to assess their experimentation activities. The committee's principles are supplemented with a mission-based approach for determining where to focus joint and naval experimentation, respectively (see the next subsection below).

[11]Such as either the appropriate Navy fleet command or the numbered fleet commands under that fleet command.

BOX 6.1
Examples of Types of Mission Areas

- *Antisubmarine warfare and antisurface warfare.* These are predominantly Navy missions, although aircraft of other Services could be used to attack the submarines and ships.
- *Antiair warfare and strike warfare.* These missions are inherently joint. The coordinated employment of assets of all Services could be used in them.
- *Operational command, control, and communications (C3) and intelligence, surveillance, and reconnaissance (ISR).* These missions are also inherently joint. The C3 will coordinate the assets of all Services, and the ISR data will be shared among all Services.
- *Logistics.* This is an evolving area. While it is primarily Service-specific now, there are opportunities for joint considerations (e.g., the sharing of maintenance and supplies).

A Mission-Based Approach for Balancing Naval and Joint Experimentation

In addition to the principles recommended by the committee, a process is necessary in order to systematically and thoroughly achieve balance between naval-specific and joint concept development and experimentation. This subsection outlines one such process for establishing mission areas that are inherently joint or Service-unique[12] for determining how experimentation can be applied to increase combat effectiveness and for prioritizing elements of the experimentation program.

The first step in this process is for the Navy and Marine Corps to review all of their missions to determine which ones each Service would largely perform in combat by itself and which ones are inherently joint. Some examples of the different types are provided in Box 6.1.

The second step in the process is to determine for each mission area how concept development and experimentation can be applied in order to increase combat effectiveness. This involves articulating current shortcomings in mission execution and identifying where Navy and Marine Corps concepts and capabilities can be brought to bear for addressing both naval-specific and joint shortcomings. Some examples of such opportunities are listed in Box 6.2.

As the last step in the process—after the missions have been laid out and the problem areas and potential opportunities for each have been identified—priori-

[12]The CFFC/NWDC draft Sea Trial Experimentation Campaign Plan appears to have followed a similar mission-based approach in constructing parts of its campaign plan.

BOX 6.2
Examples of Effectiveness Opportunities for Mission Areas

• *Antisubmarine warfare and antisurface warfare.* In these mission areas, two threats against which increased capabilities must be developed are, respectively, quiet diesel submarines and the swarm tactics of small boats in littoral areas. Addressing these threats is primarily a naval matter.

• *Antiair warfare (air defense).* Developing a joint air picture is still a challenge, involving both technical and procedural interoperability shortcomings. The Navy has worked in this area extensively and can contribute significantly to joint efforts.

• *Strike warfare (joint fires).* Attacking time-sensitive targets rapidly can involve detection of the targets by the assets of one Service and attack by those of another. In this context, the Navy has already considered the joint application of its Naval Fires Network to direct strikes.

• *Operational C3.* All Services are implementing "network-centric operations" in one sense or another. The Navy has much to offer the joint community, and at the same time should ensure that its concepts and technical capabilities and those of the other Services develop in a compatible manner.

ties will have to be established as to which problem areas and opportunities can be addressed. Those decisions can only be made by the senior leadership in the Navy and Marine Corps.

Given the committee's earlier recommendations on methods and processes, additional observations in the context of balancing naval and joint experimentation can be made:

• All opportunities for joint experimentation should be used—USJFCOM, experimentation within the Combatant Commands, and direct Service-to-Service interaction. Particular opportunities may lie in the greater use of fleet battle experiments for joint purposes and in increased direct interaction between the Navy and Marine Corps combat development centers and those of the other Services.

• As noted many times, the full spectrum of analysis, wargaming, simulation, and live experimentation should be applied. Joint cooperation can involve not only the use of these venues, but also tools for their development.

• A campaign approach will be needed, since most of the substantive problem areas will likely require a series of experimentation activities of different types to explore fully and to develop the requisite knowledge.

• Particular consideration should be given to the command and control aspects of the mission areas being addressed. Joint command and control appear critical in realizing the transformation of military capabilities.

- The newly developed capabilities (procedural and technical) should be inserted into operational forces as soon as possible. As necessary, these capabilities can be iterated upon and improved through a spiral development process.

SPECIFIC ENHANCEMENTS FOR THE NAVAL PROGRAMS OF EXPERIMENTATION

The terms of reference ask the committee to identify specific areas missing from the current program of experimentation. In two separate findings, one for each Service, the committee identified shortfalls in these programs. For the Navy, the finding focused on a lack of robust experimentation campaigns as well as on areas requiring expansion, and for the Marine Corps, the finding noted a lack of programs with long-term objectives. Below, the committee provides suggested enhancements to the naval programs of experimentation. Some of these are oriented toward a single Service, others involve both, and still others require joint experimentation.

Sea Strike

One area requiring experimentation is that of over-the-horizon, precise, time-critical strikes, including target recognition, identification, and designation for high-speed weapons with a high rate of fire.[13] Exploration of capability concepts and limitations for rapid recognition, combat identification, and the designation of targets up to 200 nm is needed, with a view to complete target detection to weapon within times that are consistent with precision ballistic short time-of-flight weapons. Experimentation should include fusing all source data from GPS-equipped multi-platform, multi-spectral sensors in combination with terrain imagery and commercial terrain-rendering products to provide common-view scenes that can be correlated with fast processors designed for the purpose.[14]

The Navy is developing extended-range guided munitions (ERGMs) for long-distance surface fire support. It is timely to experiment with how these munitions will be used at a 60-mile range from the guns in mission elements: for example, calling for fire from over the horizon, targeting, coordinating fires with air opera-

---

[13]Naval Studies Board, National Research Council. 2002. *Network-Centric Naval Forces: A Transition Strategy for Enhancing Operational Capabilities,* National Academy Press, Washington, D.C., Executive Summary; Naval Studies Board. 2002. *2002 Assessment of the Office of Naval Research's Air and Surface Weapons Technology Program*, National Academy Press, Washington, D.C.

[14]Naval Studies Board, National Research Council. 2002. *Network-Centric Naval Forces: A Transition Strategy for Enhancing Operational Capabilities,* National Academy Press, Washington, D.C., Executive Summary; Naval Studies Board. 2002. *2002 Assessment of the Office of Naval Research's Air and Surface Weapons Technology Program*, National Academy Press, Washington, D.C.

tions, and coordinating multimission operations of ships engaged in multiple activities (e.g., that may be simultaneously performing some or all of the following: antisubmarine warfare, long-range fire support, cooperative engagement capability (CEC), and mine warfare). These issues can be better understood through a series of experimentation activities, perhaps ending in an FBE with a multiship task group.

There are serious issues in flying UAVs to locate, identify, and communicate targets. Some were resolved ad hoc in Kosovo and Afghanistan, but many need further exploration. This concern can be extended to joint Navy/Marine Corps/ Air Force operations, with techniques expanding and changing as new weapons (e.g., the Navy's SM2 land attack missile and the various uninhabited combat air vehicles) begin transition. Time needs to be provided for learning how to use these new capabilities before they enter the fleet, rather than having personnel learn about them in actual war situations. Ultimately, the results would affect the Naval Fires Network as well as FORCEnet.

The development of ship-launched fire support weapons and launch and support systems will stress the current systems for reloading ship magazines beyond their capability. Especially if missiles are launched in rapid fire from Vertical Launch System (VLS) bays on guided-missile destroyers (DDGs) and guided-missile cruisers, the ships would have to retire to port for reloading. Rapid at-sea reloading capabilities for ships with VLS bays have been demonstrated but not fully implemented. There is a need for experiments with engineering developments that allow for at-sea resupply and expeditious handling of short time-of-flight strike weapon reloads.[15]

The Navy will acquire four cruise missile submarines (SSGNs), each of which can launch well over 100 missiles, many times the number that current attack submarines can launch. How will these SSGNs fit into fleet operations? By what criteria will firepower be allocated to the submarines and to surface forces? How will firepower be coordinated in the absence of extensive, large-bandwidth communication links with the submarines? These questions provide a fruitful area for experimentation. There is need for joint experimentation with the Air Force to explore the techniques of in-stride methods—and the effectiveness of these techniques—to clear the path to the beach using the Harvest Hammer technique (line charge analog using simultaneous, high-explosive charges precision-delivered from the air). Such experimentation is also recommended in the report of the Naval Studies Board on mine warfare.[16]

---

[15]Naval Studies Board, National Research Council. 1997. *Technology for the United States Navy and Marine Corps, 2000-2035: Volume 1: Overview*, National Academy Press, Washington, D.C., pp. 66-67.

[16]Naval Studies Board, National Research Council. 2001. *Naval Mine Warfare: Operational and Technical Challenges for Naval Forces,* National Academy Press, Washington, D.C.

There is a continuing need to explore Marine operations in built-up areas (MOBA). Special attention should be directed to on-scene responsiveness to surprise enemy tactics.[17] Many future scenarios may involve MOBA; in such situations, rapid combat success will minimize both military and civilian casualties. Such scenarios may well involve the use of nonlethal weapons, the TTPs for which have yet to be worked out in detail. Also, joint experimentation is needed to examine the effectiveness of various means of calling in and providing fleet-based surface and air fire support to Marines fighting in close, built-up quarters, with and without their artillery on scene. The issues to be investigated include the coordination and allocation of organic and called-in fire support. Much is already known by the Marines in this area, but new questions have surfaced as a result of lessons learned in recent conflicts and changing concepts of operation under Ship to Objective Maneuver/Operational Maneuver from the Sea (STOM/OMFTS).

Sea Shield

Joint Service simulations and exercises are needed for practicing and improving combat identification capabilities across Service and system interfaces, including handoff problems, in which many errors occur.[18] Work is going on in this area, particularly with respect to CEC in concert with Navy theater missile defense (TMD) experiments and Joint Theater Air and Missile Defense Office (JTAMDO) single integrated air picture efforts. However, combat identification of both ground and air targets remains an issue, in addition to deconfliction in cluttered littoral battle space, where air superiority does not ensure control in the eventuality of overland cruise missile and ballistic missile attacks on expeditionary forces or on the ships that support them.[19]

The Department of the Navy's technology investment program will have to develop sensors, weapons, and battle management command, control, and communications (BMC3) architectures and algorithms that are adaptive and flexible enough to allow responding to unexpected threat capabilities and characteristics. Such elements of the ballistic missile defense system should be combined into experimental systems for evaluation and refinement. The mature technologies from the program could conceivably be incorporated into future spirals of the Navy theater-wide missile defense systems.[20]

---

[17]Such exercises would have to give a "red team" free play, as well as allowing free-play responsiveness to the Marines.

[18]Naval Studies Board, National Research Council. 1996. *Navy and Marine Corps in Regional Conflict in the 21st Century,* National Academy Press, Washington, D.C., p. 10.

[19]Naval Studies Board, National Research Council. 2001. *Naval Forces' Capability for Theater Missile Defense,* National Academy Press, Washington, D.C., p. 11.

[20]Naval Studies Board, National Research Council. 2001. *Naval Forces' Capability for Theater Missile Defense,* National Academy Press, Washington, D.C., p. 12.

There is a need for joint operations experiments in cruise missile defense using networked, all-Service surveillance and track assets in order to maximize effectiveness in the littoral environment in which expeditionary forces will enter a theater. Joint experimentation should be leveraged to develop the operational concepts and technical capabilities necessary for joint missile defense operations.[21]

The application of network-centric capability to deployable undersea sensor arrays and low probability of intercept (LPI) transponders, together with autonomous unmanned underwater vehicles (UUVs), can provide the ISR to create and monitor a "single integrated undersea picture" and "sterilization" of undersea littoral waters in which asymmetric methods, cheap mines, intelligent mine fields, and quiet, small, diesel-electric submarines can otherwise deny access to U.S. naval forces.[22]

A relatively neglected area that would benefit from experimentation, both with individual submarines and with fleet units, is that of learning how the Navy would handle a breakout of opposition submarines in littoral waters, in which conventional antisubmarine warfare is especially difficult. Although some Office of Naval Research work with low frequency active (LFA) is ongoing, this is largely in the 6.2 applied research funding area.[23] Operating with ships, UUVs, and a real undersea opposing force to simulate enemy submarines would help the Navy learn how to deal with the problem before it presents itself in real hostilities.[24]

It is not too soon to plan experiments in mine countermeasures (MCM) warfare to understand how the new organic MCM systems that are in development can be used in a coordinated, integrated fashion. Also, there is a need to determine how ships such as DDG-51s, the newly planned Littoral Combat Ship, and ship commanders will function in multimission modes when mine warfare is added to the already large mission mix of air and missile defense and surface fire support.

Sea Basing

Sea basing for naval forces ashore is a radical new concept designed to avoid the sovereignty issues and delays attending the establishment of shore bases

---

[21]Naval Studies Board, National Research Council. 2001. *Naval Forces' Capability for Theater Missile Defense,* National Academy Press, Washington, D.C., p. 7.

[22]Naval Studies Board. National Research Council. 2000. *An Assessment of Undersea Weapons Science and Technology,* National Academy Press, Washington, D.C.

[23]Naval Studies Board, National Research Council. 1997. *Technology for the United States Navy and Marine Corps, 2000-2035: Volume 7: Undersea Warfare*, National Academy Press, Washington, D.C., pp. 31-33.

[24]The British fleet spent large amounts of fleet time hunting an Argentine submarine that may or may not have put to sea during the Falklands War. We should not have to spend our scarce resources that way in future conflicts.

when rapid expeditionary landings are necessary. While the Marine Corps has given a great deal of thought to the operational opportunities that sea basing will offer and to related issues, an extensive program of experimentation (and conceivably spiral development) will be required to bring capabilities to the field over time.

Sea basing will require new ship configurations to provide the ability to selectively move, repackage, and/or offload containers; to support at least limited flights of heavy, vertical-lift cargo aircraft; and to load and unload lighters and landing craft, air-cushioned (LCACs) from ships and at the shore or beach in high sea states (sea states up to and including sea state 3 and 4). It may eventually be found necessary to resort to large, movable, offshore platforms that can handle fixed-wing aircraft such as the C-130 or even the C-17. Ships involved in providing the sea base may not be able to stay near enough to shore to support the current pipeline and pumping distance of 2 miles for water and fuel. They may have to avoid or move away from shore defenses, including mines, ballistic missiles, and antiship missiles, to points over the horizon that are 25, 50, or even 100 miles from shore.

It is essential to understand logistics ship configurations with loadable containers that can be selected and moved about on demand, and that can be moved to a flight deck for air transport ashore, or moved to LCACs or other kinds of lighters for surface transport in seas that are not calm. It is important to evaluate the pluses and minuses of various ship configurations, as well as operations with and handling of the cargo within them, under various sea state conditions—this information is fundamental to the entire sea basing idea.

It is also essential to address the following critical issue—that the enemy's defenses (mines, missiles, submarines, and so on) may not have been suppressed adequately to protect a multibillion-dollar asset (the amphibious fleet) within 2 miles of a hostile coastline (the distance for pumping fuel and water under current capability), or even 25 miles out (the distance that current doctrine says is needed for protection of the fleet during the initial landing). The Marine Corps concept of operations (CONOPS) for sea basing must inevitably degrade as the required standoff distances grow from 25 miles to, say, 100 miles. Experimentation with a surrogate sea base is needed as it is positioned, say, 25, 50, and 100 miles from a coastline, to ascertain what happens to its ability to support forces of a given size at a given range inland. Will more airlift capability be required? Will it be necessary to make a sea base large enough to accommodate fixed-wing aircraft such as a C-17? The experimentation campaign plan for forcible entry from a sea base should address the need for and explore the feasibility of new CONOPS, architectures that are not dependent on the advanced, amphibious, assault vehicle (AAAV), the Osprey tiltrotor (the V-22), or other systems having large daily logistics requirements.

At some point the Navy and Marine Corps will have to move large, heavy things over the shore from lighters and LCACs, or set up over-the-shore pipelines—water and fuel are among the biggest logistics items. There has not been a great deal of experimentation in this area, especially in overcoming the sea state 3 barrier. This area is ripe for extensive experimentation, spiraling toward viable concepts and capabilities.

Another area for experimentation is that of joint Navy and Marine Corps cooperation involving a landing over the shore in OMFTS/STOM mode, with simulated firepower support. There are problems in exercising force-wide command and control during such a landing; and in sustaining the movement and its support from the sea base for a week, or even a month, of uninterrupted operations. The results of experimentation on logistics ship configurations are also relevant to this area of investigation.

The sea-basing issues discussed above, including advanced support and logistics concepts for OMFTS and Sea Basing, would benefit greatly from extensive study through a spectrum of experimentation activities, among them modeling and simulation of these operations, to assess logistics flow needs, capabilities, and alternative support concepts.[25] Ultimately, experimentation at the ship and multiship/air wing level will be needed to provide answers to such questions as the ones posed above.

The Marine Corps is moving in the direction of adapting the network-centric concept to delivery-as-needed logistics management. Because of the complexity of these issues, at some point the modeling and simulation that supports the concept development will also require validation as part of joint experiments and field events to assure that the simulation models are matched by reality.

---

[25]Naval Studies Board, National Research Council. 1999. *Naval Expeditionary Logistics: Enabling Operational Maneuver from the Sea,* National Academy Press, Washington, D.C., p. 10.

Appendixes

A

Biographies of Committee Members and Staff

Annette J. Krygiel (Chair) is an independent consultant, having recently completed an assignment as a visiting fellow at the Institute for National Strategic Studies at the National Defense University, where she wrote a book on large-scale system integration. Dr. Krygiel's background is in the management of large-scale systems, particularly in regard to software development and systems integration. Before being appointed to the Institute for National Strategic Studies, she was director of the Central Imagery Office (CIO), a Department of Defense combat support agency. Dr. Krygiel remained the director for 27 months, until CIO joined the National Imagery and Mapping Agency in October 1996. Dr. Krygiel began her career at the Defense Mapping Agency, where she held various positions, such as chief scientist. Dr. Krygiel has been a participant in National Research Council (NRC) studies, including the work of the Panel on Distributed Geolibraries: Spatial Information Resources. She is currently a member of the NRC's Naval Studies Board.

Ruzena K. Bajcsy is director of the Center for Information Technology Research in the Interest of Society (CITRIS) at the University of California, Berkeley. Before becoming director for CITRIS, she was assistant director for Computer Information Science and Engineering at the National Science Foundation. A member of the National Academy of Engineering and Institute of Medicine, Dr. Bajcsy obtained her first Ph.D. from Slovak Technical University (Slovakia) and her second Ph.D. from Stanford University. Dr. Bajcsy's research interests are in machine perception, computer vision, characterizing and solving problems involving segmentation, and three-dimensional vision and other sensory modalities that function together with vision. She has served on numerous scientific and technical advisory groups including the NRC's Army Research Laboratory Technical Assessment Board. She is a fellow of the Institute of Electrical and Electronics Engineers (IEEE) and founding fellow of the American Association of Artificial Intelligence.

Alan Berman is a part-time employee at the Applied Research Laboratory of Pennsylvania State University (ARL/PSU) and at the Center for Naval Analyses (CNA). At ARL/PSU, Dr. Berman provides general management support and program appraisal; at CNA, he assists with analyses of Navy research and development investment programs, space operation capabilities, information operations, and command, control, communications, computers, intelligence, surveillance, and reconnaissance (C4ISR) programs. Dr. Berman's background is in defense science and technology, particularly in regard to advanced weapons and combat systems. His previous positions include that of dean of the Rosenstiel School of Marine and Atmospheric Sciences at the University of Miami, where he was responsible for the graduate programs in physical oceanography, marine biology, geology, geophysics, applied ocean science, and underwater acoustics; and director of research at the Naval Research Laboratory, where he administered broad programs in basic and applied research, including electronic warfare, radar, communications, space systems, space sciences, material sciences, plasma physics, antisubmarine warfare, underwater acoustics, oceanography, electronic devices, and space-based time standards for Global Positioning Systems. Dr. Berman has served on numerous scientific boards and advisory committees and is currently a special adviser to the NRC's Naval Studies Board.

Duncan A. Brown is a principal staff member at the Applied Physics Laboratory of Johns Hopkins University (APL/JHU). His responsibilities include strategic and business planning, wartime assessments oversight, and the creation and execution of a Johns Hopkins National Security Studies and Technology Seminar. Before joining APL/JHU, Mr. Brown served as the principal research and development adviser to the Deputy Chief of Naval Operations for Requirements and to the Commander in Chief of the Pacific Fleet. He received an M.S. in technical management from Johns Hopkins University and an M.S. in ocean engineering from the University of Rhode Island.

Marion R. Bryson is director of research and development for North Tree Management, an entrepreneurial activity in Monterey, California. He has held many positions in the federal government, spending 22 years primarily in the operational test arena. He served as scientific adviser at the U.S. Army Combat Development Experimentation Command (CDEC), as director of CDEC, and as technical director of the Test and Experimentation Command. Prior to his government service, Dr. Bryson taught in several colleges and universities, including Duke University. He is past-president of the Military Operations Research Society and recently served on the NRC Panel on Statistical Methods for Testing and Evaluation.

John D. Christie is a senior fellow at the Logistics Management Institute. He has an extensive background in Department of Defense acquisition policy and program analysis. From 1989 to 1993, Dr. Christie served as director of Acquisition Policy and Program Integration for the Office of the Under Secretary of Defense (Acquisition). In that role he directed the preparation of a comprehensive revision to all defense acquisition policies and procedures resulting in the cancellation and consolidation of 500 prior separate issuances. (He also prepared comprehensive acquisition program alternatives for the Secretary of Defense that resulted in multibillion-dollar budget reductions.) As a former member of Army Science Board, Dr. Christie was called upon to direct reviews of the Army analytical community and operations research activities for the Vice Chief of Staff, including the support of the overall Army acquisition process and its integration with the programming and budgeting process. A member of the NRC's Naval Studies Board, he has served on numerous scientific boards and advisory committees, including the Commission on Roles and Missions of the Armed Forces that provided recommendations to improve defense management.

John A. Corder retired from the U.S. Air Force as a Major General in 1992. General Corder's background is in U.S. Air Force operational and joint issues. Since his retirement, he has been employed as an independent technical adviser. He has served as an ad hoc adviser to the Air Force Scientific Advisory Board on the subjects of theater battle management, theater air defense, and tactical ballistic missile defense. General Corder is a former command pilot and navigator; he flew more than 100 combat missions over North Vietnam. His military career also included assignments as director of Electronic Combat and commander of the 65th Electronic Combat Air Division with the U.S. Air Force in Europe. He was deputy commander for Air Combat Operations for the Central Command Air Forces in the 1991 Persian Gulf War. General Corder was responsible for the planning and execution of 3,000 combat sorties per day—an effort that involved the coordination of Air Force, Navy, Marine Corps, and Allied aircraft from nine other nations.

Paul K. Davis is professor of policy analysis at the RAND Graduate School of Policy Studies and senior scientist and research leader at RAND, where his research interests include defense planning, reengineering and transformation of the U.S. military deterrence theory, new analytic methods and modeling, and strategic planning generally. He has published more than 100 papers and books on these and other subjects. During his tenure at RAND, Dr. Davis has served tours as corporate research manager for defense and technology planning and program manager for strategy planning and assessment. Before joining RAND, he was a senior executive in the Office of the Secretary of Defense (Program Analysis and Evaluation). Dr. Davis has served on numerous scientific boards and advisory committees and is currently a member of the NRC's Naval Studies Board.

Brig "Chip" Elliott is principal engineer for BBN Technologies. There he has led the design and successful implementation of a number of secure, mission-critical networks based on novel Internet technology for the United States, Canada, and the United Kingdom (Near-Term Digital Radio, Joint Tactical Radio System, Iris, Bowman) and has acted as senior adviser on a number of national and commercial networks including three low-Earth-orbit satellite constellations (Discoverer II, Space-Based Infrared Surveillance Low, and Celestri/Teledesic) and Boeing's Connexion system. Mr. Elliott has particular expertise in wireless Internet technology, mobile ad hoc networks, quality-of-service issues, and novel routing techniques. At present he is leading the design and build-out of a very highly secure network protected by quantum cryptography. He holds more than 70 patents pending or issued on network technology, currently serves on the NRC's Naval Studies Board, and has participated in a variety of other national advisory panels including the Defense Science Board and Army Science Board.

J. Dexter Fletcher is a member of the senior research staff at the Institute for Defense Analyses. A psychologist by training, he is a leading authority on military education and training issues, particularly on aspects related to the use of advanced technologies and human-machine interface. Dr. Fletcher's research interests include knowledge management, online curriculum development, and virtual reality. He received an M.S. in computer science and a Ph.D. in educational psychology from Stanford University.

Richard J. Ivanetich is director of the Computer and Software Engineering Division at the Institute for Defense Analyses (IDA). Before assuming that position in 1990, he was assistant director of the System Evaluation Division at IDA. Dr. Ivanetich's research interests are in the areas of defense systems, technology, and operations analyses, and he is primarily concerned with computer and information systems; command, control, and communications systems and procedures, modeling and simulation of systems and forces, crisis management, and strategic and theater nuclear forces. Before joining IDA in 1975, Dr. Ivanetich was assistant professor of physics at Harvard University. He is a member of the NRC's Naval Studies Board.

L. David Montague, an independent consultant, is retired president of the Missile Systems Division at Lockheed Martin Missiles and Space and a former officer of Lockheed Corporation. A member of the National Academy of Engineering (NAE), he has more than 40 years of experience in the design, development, and program management of military weapons and their related systems. His expertise includes complex systems engineering and systems integration of ballistic missiles, cruise missiles, and unmanned aerial vehicles, as well as exo- and endoatmospheric interceptors engaging these classes of threats. In addition, he is knowledgeable about threat assessment; overhead surveillance systems; cueing technology; battle and engagement management methodology; interceptor design, guidance and control, countermeasures, and discrimination; and the use of directed-energy weapons for defensive purposes. Mr. Montague is a fellow of the American Institute of Aeronautics and Astronautics (AIAA) and a previous recipient of the AIAA's Missile Systems Award. He has served on numerous scientific boards and advisory committees, including the Navy Strategic Systems Steering Task Group and task forces for both the U.S. Army and the Defense Science Board.

William B. Morgan recently retired after having spent his entire professional career at the Naval Surface Warfare Center, Carderock Division (formerly the David Taylor Research Center). He retired as head of the Hydromechanics Directorate. From 1951 to 1958, he served as a naval architect in the Propeller Branch. From 1958 to 1962, he served as supervisor of the research and design section of the Propeller Branch. From 1962 to 1970, he served as head of the Propeller Branch. In 1970, he became head of the Naval Hydromechanics Division, and in December 1979, head of the directorate. Dr. Morgan has made significant contributions to the field of naval architecture during his long and distinguished career. He has authored and coauthored numerous papers and reports dealing with a wide variety of subjects in the propulsion area. A member of the National Academy of Engineering, he has served on numerous scientific boards and advisory committees and is currently a member of the NRC's Naval Studies Board.

Jason Providakes is executive director of the Robust Global Operations Program at MITRE Corporation. He joined MITRE in 1989 as lead scientist; there, his research interests included areas relating to over-the-horizon radar, integrated optics, and command, control, and communications (C3). During his tenure at MITRE, Dr. Providakes has served in key technical roles within the Air Force C3 Center and the Washington C3 Center. He recently led efforts for the Air Force on architecture development, system engineering, and integration activities. Before joining MITRE, Dr. Providakes did graduate work in the Electrical Engineering Department at Cornell University, where he obtained a Ph.D. degree in 1985; from 1985 through 1989, he taught and performed research in the area of radar backscatter to study Earth's ionosphere.

John E. Rhodes retired as Lieutenant General from the U.S. Marine Corps in August 2000. General Rhodes's last position was as Commanding General of the U.S. Marine Corps Combat Development Command, where he led the Marine Corps in its efforts to develop warfighting concepts and integration of all aspects of doctrine, organization, training, materiel, leadership, personnel, and facilities to enable the Marine Corps to field combat-ready forces. This responsibility entailed, among other things, careful assessments of current and future operating environments and continuous adaptation of the Marine Corps's training infrastructure and resources in order to ensure that the integrated capabilities were continuously developed for the Unified Commanders in Chief.

William D. Smith retired as Admiral from the U.S. Navy in 1993 after 38 years of active duty service. Admiral Smith's background is in Navy planning, programming, budgeting, and operational issues. His last assignment was as U.S.

Military Representative to the NATO Military Committee in Brussels, Belgium. Admiral Smith has served in a number of high-ranking capacities for the Chief of Naval Operations. From 1987 to 1991, he served as Deputy Chief of Naval Operations for Logistics and Navy Program Planning. From 1985 to 1987, he was director, Fiscal Management Division/Comptroller of the Navy. He is currently a member of the NRC's Naval Studies Board.

Michael G. Sovereign is professor emeritus of command, control, and communications (C3) at the Naval Postgraduate School. Dr. Sovereign's background is in C3, joint warfare analysis, and acquisition cost-cycle analysis. He was formerly the director of the Institute of Joint Warfare and served as visiting research professor for Headquarters, U.S. Commander in Chief Pacific Fleet, where his responsibilities included conducting research on the Navy's Virtual Information Center workshops and other experiments aimed at addressing joint C4ISR issues. Dr. Sovereign was also senior principal scientist at the Supreme Headquarters Allied Powers, Europe (SHAPE) Technical Center (now, NATO C3 Agency), where he participated in major replanning of NATO C3 systems, and once served as director of special projects in the Office of the Assistant Secretary of Defense (Comptroller), where he directed the revision of the Department of Defense's planning, programming, and budgeting system and instituted methods for measuring and budgeting for inflation in weapons systems. Dr. Sovereign has authored numerous articles on instructional media, defense logistics, and economics.

Mitzi M. Wertheim is a consultant to Enterprise Solutions at the Center for Naval Analyses (CNA); her expertise there is in the application of business process reengineering methods and teaching large corporations to increase service while reducing cost. In recent years, her research interests have focused on naval career, education, and training issues. Before joining CNA, Ms. Wertheim was vice president of Enterprise Solutions at SRA International, Incorporated. At SRA, she creatively applied business process reengineering methods in order to improve productivity and lead to higher levels of service while reducing the costs to large corporations of becoming customer-focused, process-focused, and team-oriented using information technology (IT) as an enabler. Before joining SRA, Ms. Wertheim worked with IBM Federal Systems Company as an enterprise consultant, marketing manager, program manger, and technical assistant. From 1977 to 1981, she was the Deputy Undersecretary of the Navy. A member of the NRC's Naval Studies Board, Ms. Wertheim has served on numerous committees and scientific advisory boards, including the Council on Foreign Relations and the Advisory Board of the Defense Budget Group.

Cindy Williams is a principal research scientist of the Security Studies Program at the Massachusetts Institute of Technology. Formerly she was assistant director for national security at the Congressional Budget Office, where she led the National Security Division in studies of budgetary and policy choices related to defense and international security. Dr. Williams has served as a director and in other capacities at the MITRE Corporation in Bedford, Massachusetts; as a member of the Senior Executive Service in the Office of the Secretary of Defense at the Pentagon; and at RAND in Santa Monica, California. She is the editor and one of several authors of *Holding the Line: U.S. Defense Alternatives for the Early 21st Century*, a book about defense spending choices for the future. Her areas of specialization include the national security budget, command and control of military forces, conventional air and ground forces, and nuclear weapons.

Joseph Zeidner is professor emeritus of administrative sciences and psychology at the Graduate School of Arts and Sciences, George Washington University (GWU). Dr. Zeidner's background is in human resources and the factors involved in learning, training, decision making, and job performance. He has written several books on the economic benefits of predicting job performance and estimating the gains of alternative policies in affecting human performance. He has been influential in personnel classification issues and contributed to the *Encyclopedia of Human Intelligence*. Beforejoining GWU in 1982, Dr. Zeidner served as the technical director of the U.S. Army Research Institute and as chief psychologist of the U.S. Army.

Staff

Charles F. Draper is acting director at the National Research Council's Naval Studies Board. He joined the National Research Council in 1997 as program officer, then senior program officer, with the Naval Studies Board and in 2003 became associate director. During his tenure with the Naval Studies Board, Dr. Draper has served as the responsible staff officer on a wide range of topics aimed at helping the Department of the Navy with its scientific, technical, and strategic planning. His recent efforts include topics on network-centric operations, theater missile defense, mine warfare, and nonlethal weapons. Prior to joining the Naval Studies Board, he was the lead mechanical engineer at Sensytech, Inc. (formerly S.T. Research Corporation), where he provided technical and program management support for satellite Earth station and small-satellite design. He received his Ph.D. in mechanical engineering from Vanderbilt University in 1995; his doctoral research was conducted at the Naval Research Laboratory (NRL), where he used an atomic force microscope to measure the nanomechanical properties of thin-film materials. In parallel with his graduate student duties, Dr. Draper was a mechanical engineer with Geo-Centers, Inc., working on-site at NRL on the development of an underwater x-ray backscattering tomography system used for the nondestructive evaluation of U.S. Navy sonar domes on surface ships.

Seymour J. Deitchman is a private consultant on national security, research and development management, and systems evaluation. Mr. Deitchman, a mechanical and aeronautical engineer by training, served at the Institute for Defense Analyses (IDA) as vice president for programs, with responsibility for planning and supervising the IDA research program and ensuring its quality and performance. He was on occasion delegated responsibility to be CEO in the absence of the president of IDA. As a senior research associate subsequent to his retirement from IDA management, he was responsible for projects on tactical aviation, man in space, and man-in-the loop simulation. Before becoming vice president for programs he held other positions at IDA, including assistant vice president for research and vice president for planning and evaluation. As director of the Office of Civil Programs at IDA, he dealt with transportation and with environment and law enforcement issues and systems. Mr. Deitchman was director of Overseas Defense Research at the Advanced Research Projects Agency (ARPA), where he was responsible for planning and executing ARPA's R&D program on counterinsurgency and related technical matters in support of U.S. military operations in Southeast Asia and some operations in the Middle East. In the Office of the Director of Defense Research and Engineering, he was special assistant for counterinsurgency. In this position he established the DOD program of R&D support for Southeast Asia operations. Before entering the Department of Defense, Mr. Deitchman held various positions, as an IDA staff member, with the Cornell Aeronautical Laboratory (now CALSPAN Corporation), at the Bell Aircraft Corporation, and at the National Advisory Committee for Aeronautics (now NASA), with responsibility for programs having to do with limited war, tactical aviation, Army air mobility, air traffic control, aircraft aerodynamic design, and wind tunnel testing. He has been a member of various government advisory panels, an occasional lecturer at the National War College and Industrial College of the Armed Forces, and a U.S. Delegate to NATO Defense Research Group and the Advisory Group on Aeronautical Research and Development. He is the author of five books and numerous published papers on national security matters.

B

Agendas for Meetings of the Committee

APRIL 4-5, 2002
NATIONAL RESEARCH COUNCIL, WASHINGTON, D.C.

Thursday, April 4, 2002

Closed Session: Committee Members and NRC Staff Only

0800 CONVENE—Welcome, Introductions
Dr. Annette Krygiel, Committee Chair
Dr. Ronald Taylor, Director, Naval Studies Board (NSB)

0900 COMMITTEE DISCUSSION
Moderator: Dr. Annette Krygiel, Committee Chair

Data-Gathering Meeting Not Open to the Public: Classified Discussion (Secret)

The Role of Experimentation in Building Future Naval Forces to Operate in the Joint Environment; Military Experimentation Overview and Progress Summary; Supporting Science and Technology

1000 NAVAL TRANSFORMATION PLAN—Overview, Future Forces
RADM Daniel R. Bowler, USN, Director, Warfare Integration and Assessment, Office of the Chief of Naval Operations (OPNAV), N70

1030 EXPERIMENTATION OVERVIEW—The Naval Transformation Plan and How Experimentation Is Organized and Managed
CAPT David W. Gillard, USN, Branch Head, Technology, Experimentation, Transformation and Doctrine for Warfare Integration and Assessment, OPNAV N70T

1230 JOINT EXPERIMENTATION OVERVIEW AND PROGRESS SUMMARY—Comprehensive Summary of Military Experimentation; How Experimentation Is Organized and Managed; Progress and Examples
Mr. John A. Klevecz, Director, Futures Alliance, Joint Experimentation, U.S. Joint Forces Command (USJFCOM), J9

1530 EXPERIMENTATION S&T OVERVIEW AND PROGRAM SUMMARY—Description of Naval S&T Programs; Implementation, Progress and Examples
Mr. Paul M. Lowell, Chief of Staff, Office of Naval Research (ONR)

Closed Session: Committee Members and NRC Staff Only

1700 COMMITTEE DISCUSSION
Moderator: Dr. Annette Krygiel, Committee Chair

1830 END SESSION

Friday, April 5, 2002

Closed Session: Committee Members and NRC Staff Only

0800 CONVENE—Welcome, Committee Discussion
Dr. Annette Krygiel, Committee Chair
Dr. Charles Draper, Sr. Program Officer, NSB

Data-Gathering Meeting Not Open to the Public: Classified Discussion (Secret)

The Role of Experimentation in Building Future Naval Forces; Previous Studies and Reviews

0900 MARINE CORPS EXPERIMENTATION AND OVERVIEW—How USMC Experimentation Is Organized and Managed
Mr. Hugh Montgomery, Technical Director, Marine Corps Warfighting Laboratory (MCWL)
Mr. Fred Belen, Division Director, Expeditionary Warfare Operations Technology Division, ONR

1045 NAVY EXPERIMENTATION AND OVERVIEW—How USN Experimentation Is Organized and Managed
RADM Robert G. Sprigg, USN, Commander, Navy Warfare Development Command (NWDC)

1300 PREVIOUS STUDIES AND REVIEWS
Dr. David S. Alberts, Director, Research and Strategic Planning, Office of the Assistant Secretary of Defense for Command, Control, Communications and Intelligence (OASD C3I)

Closed Session: Committee Members and NRC Staff Only

1500 COMMITTEE DISCUSSION
Moderator: Dr. Annette Krygiel, Committee Chair

1700 ADJOURN

MAY 2-3, 2000
NATIONAL RESEARCH COUNCIL, WASHINGTON, D.C.

Thursday, May 2, 2002

Closed Session: Committee Members and NRC Staff Only

0800 CONVENE—Welcome, Introductions, Plans for the Day, Taxonomy Discussion
Moderator: Dr. Annette Krygiel, Committee Chair

Data-Gathering Meeting Not Open to the Public: Classified Discussion (Secret)

0900 NAVY EDUCATION AND TRAINING IN THE CONTEXT OF EXPERIMENTATION
Dr. Allen R. Zeman, Director, Naval Training and Education, Office of the Chief of Naval Operations, N79

1030 MARINE CORPS SYSTEMS ENGINEERING EFFORTS AND THE SYSCOM'S ROLE IN SPIRAL DEVELOPMENT
BrigGen James M. Feigley, USMC, Commander, Marine Corps Systems Command

1300 MARINE CORPS CONCEPT, DOCTRINE DEVELOPMENT AND EDUCATION, TRAINING EFFORTS
BrigGen Kenneth J. Glueck, Jr., USMC, Director, Warfighting Development Integration Division, Marine Corps Combat Development Command

1430 PERSPECTIVE ON RUNNING A LARGE-SCALE EXPERIMENT (EXTENDING THE LITTORAL BATTLESPACE)
Mr. Frederick E. Belen, Director, Expeditionary Warfare Operations Technology Division, Office of Naval Research

Closed Session: Committee Members and NRC Staff Only

1600 COMMITTEE DISCUSSION—Taxonomy Discussion (Continued)
Moderator: Dr. Annette Krygiel, Committee Chair

1700 END SESSION

Friday, May 3, 2002

Closed Session: Committee Members and NRC Staff Only

0800 CONVENE—Welcome, Plans for the Day, Committee Discussion
Moderator: Dr. Annette Krygiel, Committee Chair

Data-Gathering Meeting Not Open to the Public: Classified Discussion (Secret)

0900 NAVY SYSTEMS ENGINEERING EFFORTS AND THE SYSCOM'S ROLE IN SPIRAL DEVELOPMENT
VADM George P. Nanos, USN, Commander, Naval Sea Systems Command

1030 STATUS OF ASN RDA CHIEF ENGINEER
RADM Michael G. Mathis, USN, Assistant Deputy Commander for Surface Ship Technology, Naval Sea Systems Command; Commander, Naval Surface Warfare Center; Chief Engineer of the Navy; and Single Integrated Air Picture Systems Engineer (SIAP/SE), Joint Services Program
Mr. Michael J. O'Driscoll, Deputy Chief Engineer of the Navy, Office of the Assistant Secretary of the Navy for Research, Development, and Acquisition

1230 PERSPECTIVE ON NAVY EXPERIMENTATION
VADM Herbert Browne, USN (Ret.), President, Armed Forces Communications and Electronics Association

1400 PERSPECTIVE ON ARMY EXPERIMENTATION
GEN William W. Hartzog, USA (Ret.), President and COO, Burdeshaw Associates

Closed Session: Committee Members and NRC Staff Only

1530 COMMITTEE DISCUSSION—Wrap-Up, Meeting Reflection, Plans for Next Meetings
Moderator: Dr. Annette J. Krygiel, Committee Chair

1600 ADJOURN

JULY 9-11, 2002
NAVY WARFARE DEVELOPMENT COMMAND, NEWPORT, RI

Tuesday, July 9, 2002

0815 OPENING REMARKS/WELCOME

Data-Gathering Meeting Not Open to the Public: Classified Discussion (Secret)

0830 NWDC COMMAND BRIEF—Overview of command mission, functions, manning and products

0930 CONCEPT DEVELOPMENT BRIEF —Discussion of how concepts are developed at NWDC focusing on the five key elements: information knowledge advantage, assured access, effects based operations, anti-terrorism/force protection, and forward sea-based logistics
CAPT John Meyer, USN, Dept. Head

1030 TECHNOLOGY DIRECTOR BRIEF—Discussion of technologies that will impact the future of the Navy
Mr. Wayne Perras, Technology Director

1330 OPERATIONS DEPARTMENT BRIEF—Discussion of the areas functionally supporting experimentation at NWDC
CAPT Richard Medley, USN, Department Head

1430 MARITIME BATTLE CENTER BRIEF—Discussion of the logistics and execution of fleet battle experiments
CAPT Patrick Denny, USN, Department Head

1540 MODELING AND SIMULATION BRIEF—Presentation on how the experimental environment is built and communications, systems integration, modeling, and simulation

1830 END SESSION

Wednesday, July 10, 2002

Data-Gathering Meeting Not Open to the Public: Classified Discussion (Secret)

0815 OPENING REMARKS/OVERVIEW COMMENTS REGARDING DAY TWO BRIEFINGS

0830 DOCTRINE DEPARTMENT BRIEF—Discussion of the documentation of current fleet doctrine and experimentation results implemented in the fleet

0930 ASSURED ACCESS WIDT BRIEF—Detailed discussion regarding AA development and future direction

1030 FORWARD SEA BASED FORCES WIDT BRIEF—Detailed discussion regarding FSBF development and future direction

1300 ANTI-TERRORISM/FORCE PROTECTION WIDT BRIEF—Discussion regarding AT/FP & HLS development and future direction

1400 Information Knowledge Advantage WIDT Brief—Detailed discussion regarding IKA development and future direction

1450 Break/Refreshments

1505 ForceNet Brief (CNO Strategic Studies Group)

1615 Visit Naval War College

1730 End Session

Thursday, July 11, 2002

Data-Gathering Meeting Not Open to the Public: Classified Discussion (Secret)

0830 Effects Based Operation WIDT Brief—Detailed discussion regarding EBO development and future direction

1000 Maritime Battle Center Brief (Cont.)
CAPT Patrick Denny, USN, Dept. Head

1115 Wrap-Up/Roundtable Discussion
Moderator: Dr. Annette Krygiel, Committee Chair

1200 Adjourn

JULY 30-AUGUST 1, 2002
U.S. JOINT FORCES COMMAND AND NORFOLK NAVAL STATION
HAMPTON ROADS, VA

Tuesday, July 30, 2002

Data-Gathering Meeting Not Open to the Public: Classified Discussion (Secret)

0915 Opening Remarks
MajGen Leo V. Williams III, USMC Reserve, Vice Director, Joint Experimentation, USJFCOM
Mr. John A. Klevecz, Director, Futures Alliance, Joint Experimentation, USJFCOM

0930 U.S. Joint Forces Command Overview—Command Mission, Functions, Manning, and Products; Summary of Joint Experimentation to Date (e.g., Major Experiments, Lessons Learned in Experiment Planning, Execution, and Transition of Results, Insights into Leaps in Transformation); Way Ahead for USJFCOM
VADM Martin J. Mayer, USN, Deputy Commander in Chief, USJFCOM

1045 U.S. JOINT FORCES COMMAND EXPERIMENTATION PROCESS—Overview of Joint Experimentation Process (e.g., Establishment of Plans, Setting of Priorities, Balance Between Large-Scale and Smaller Experiments, Coordination with Services and Combatant Commanders, Transition of Results); Concept Development and Overview of Key Concepts (e.g., Rapid Decisive Operations, Effects-Based Operations, Standing Joint Force Headquarters); Environments and Tools Supporting Joint Experimentation (e.g., Simulation, Analysis, System Integration)
Mr. John A. Klevecz, Director, Futures Alliance, Joint Experimentation, USJFCOM

1230 MILLENNIUM CHALLENGE '02—Experiment/Exercise Design, Extent of Planning and Preparation, Preliminary Results; Observation of Live Feeds Between the Joint HQ and the Service Operational Forces in the Western Range Training Complex; Interactive Session with the National Elements of Power; Demonstration of the Common Relevant Operational Picture; and In-Depth Orientation on the Joint Enroute Mission Planning and Rehearsal System
CAPT Richard A. Feckler, USN, J9239

1345 MILLENNIUM CHALLENGE '02—TURNING CONCEPTS INTO CAPABILITIES (WAY AHEAD)
LTC Kevin M. Woods, USA

1400 FLEET PERSPECTIVE ON NAVAL AND JOINT EXPERIMENTATION
Classified VTC to VADM James C. Dawson, Jr., USN, Commander Second Fleet/Commander, Striking Fleet Atlantic

1515 TRAVEL TO JOINT WARFARE FIGHTING CENTER (JWFC)
Greeted by MajGen Gordon C. Nash, USMC, Commander, Joint Warfighting Center

1530 MILLENNIUM CHALLENGE '02, DEMONSTRATIONS, TOURS—TEST BAY 31—Range Integration Demonstration, Operational Net Assessment; TEST BAY 15—Joint Exercise Control Group (JECG) Update; and TEST BAY 14—JOC/MC '02 Update, Joint Task Force Perspective

1730 MILLENNIUM CHALLENGE '02—Transformation Over Time; Standing Joint Force Headquarters; Millennium Challenge Behind the Scenes; Joint Forces in the Future; Service Videos; Joint Public Affairs Operations Group; Joint Enroute Mission Planning and Rehearsal System (Near Term Video); Range Integration/Joint National Training Capability
Brig Gen James B. Smith, USAF, Vice Commander, Joint Warfighting Center and Deputy Director, Joint Training

1800 END SESSION

Wednesday, July 31, 2002

Data-Gathering Meeting Not Open to the Public: Classified Discussion (Secret)

0830 SUBMARINE FORCE, U.S. ATLANTIC FLEET OVERVIEW AND PERSPECTIVE ON EXPERIMENTATION CONTEXT, AND CONCEPT DEVELOPMENT AND EVALUATION

Mr. Floyd D. (Ken) Kennedy, Jr., NAVSUBFOR N02EG/SUBLANT N7EG, Special Assistant for Concept Development and Experimentation, Center for Naval Analyses Representative

TACTICAL DEVELOPMENT AND EVALUATION

Commodore Scott Van Buskirk, USN, COMSUBDEVRON 12 (CSDS 12)

0945 TRAVEL TO LANTFLT COMPOUND (BUILDING NH1, FLAG BRIEFING ROOM)

1000 U.S. ATLANTIC FLEET/U.S. FLEET FORCES PERSPECTIVE ON NAVAL AND JOINT EXPERIMENTATION

VADM Albert H. Konetzni Jr., USN, Deputy Commander in Chief/Chief of Staff, U.S. Atlantic Fleet

1130 TRAVEL TO OPTEVFOR (MAIN CONFERENCE ROOM)

1145 OPERATIONAL TEST AND EVALUATION FORCE OVERVIEW OF RESPONSIBILITIES AND PERSPECTIVES ON EXPERIMENTATION

RADM David M. Crocker, USN, Special Assistant to Commander in Chief, U.S. Atlantic Fleet/Commander, Operational Test and Evaluation Force

1330 U.S. AIR FORCE EXPERIMENTATION EFFORTS—Environments and Tools Supporting USAF Experimentation (e.g., Combined Air Operations Center-Experimental (CAOC-X (C2)), the C2 Battlelab, the Innovation Division); Demonstrations

Wing Commander Reginald Carey, RAF, Deputy Chief, Air Operations Support Division, USAF Air and Space C2 and ISR Center (AFC2ISRC)

1515 TRAVEL TO U.S. AIR FORCE AIR COMBAT COMMAND (BUILDING 204, MAIN CONFERENCE ROOM)

1530 TRANSITIONING U.S. AIR FORCE REQUIREMENTS INTO ACQUISITION THROUGH EXPERIMENTATION

Brig Gen Joseph P. Stein, USAF, Director of Requirements, USAF Air Combat Command

Col Gregory A. Feest, USAF, Deputy Director of Requirements, USAF Air Combat Command

1700 END SESSION

Thursday, August 1, 2002

Data-Gathering Meeting Not Open to the Public: Classified Discussion (Secret)

0830 COMMANDER, CARRIER GROUP FOUR/COMMANDER, CARRIER STRIKING FORCE PERSPECTIVE ON EXPERIMENTATION AND TRAINING OF ALL EAST COAST BATTLE GROUPS

RADM Lindell G. Rutherford, USN, Commander, Carrier Group Four/Commander, Carrier Striking Force

0930 TRAVEL TO NAVAL AMPHIBIOUS BASE (NAB) LITTLE CREEK (BUILDING 1265, COMMAND BRIEFING ROOM)

1015 NAVAL NETWORK WARFARE COMMAND OVERVIEW AND ROLE IN EXPERIMENTATION

VADM Richard W. Mayo, USN, Commander, Naval Network Warfare Command

1130 TRAVEL TO EXPEDITIONARY WARFARE TRAINING GROUP ATLANTIC (EWTGLANT)

1145 WELCOME

Col William M. Meade, USMC, Commanding Officer, EWTGLANT

RECENT EXPERIENCES IN AND AROUND KANDAHAR AND IF/HOW EXPERIMENTATION MIGHT HAVE PLAYED

Classified Teleconference to MajGen (Sel) James N. Mattis, USMC, Deputy Command General, 1st Marine Expeditionary Force/Command General, 1st Marine Expeditionary Brigade

1330 ADJOURN

AUGUST 15-16, 2002
NATIONAL ACADEMIES, WASHINGTON, D.C.

Thursday, August 15, 2002

Data-Gathering Meeting Not Open to the Public: Classified Discussion (Secret)

0745 CONVENE—Welcome, Plans for the Day, Questions on Report Status

Dr. Annette J. Krygiel, Committee Chair

Dr. Charles F. Draper, NSB Senior Program Officer

Data-Gathering Meeting Not Open to the Public: Classified Discussion (Secret)

0800 OVERVIEW OF U.S. MARINE CORPS EXPERIMENTATION EFFORTS

OPENING REMARKS AND OVERVIEW OF USMC EXPEDITIONARY FORCE DEVELOPMENT SYSTEM (EFDS)

Col James N. Strock, USMC (Ret.), Deputy Director, EFDC, Marine Corps Combat Development Command (MCCDC)

TRANSFORMATION AND CONCEPTS

Col Arthur J. Corbett, USMC, Director, Futures/Concepts, EFDC, MCCDC

JOINT CONCEPT DEVELOPMENT EXPERIMENTATION (JCDE)

Col Frank J. DiFalco, USMC, JCDE Operations Center, Warfighting Development Integration Directorate, MCCDC

MARINE CORPS WARFIGHTING LAB

Col Barry M. Ford, USMC, Chief of Staff, Marine Corps Warfighting Lab (MCWL), MCCDC

1330 RECENT ASSESSMENTS OF NAVAL EXPERIMENTATION—Strategic Framework and Experimentation Campaign Plan for Navy/Marine Corps Experimentation; Prescription for Future Navy Experimentation.

Mr. Roy Evans, Chief Engineer, The MITRE Corporation

1530 INSIGHTS INTO U.S. ARMY EXPERIMENTATION EFFORTS

MG Steven Boutelle, USA, Director of Information Operations, Networks, and Space, DISC4

Closed Session: Committee Members and NRC Staff Only

1700 WRAP-UP/COMMITTEE DISCUSSION

Moderator: Dr. Annette J. Krygiel, Committee Chair

1730 END SESSION

Friday, August 16, 2002

Closed Session: Committee Members and NRC Staff Only

0800 CONVENE—Committee Deliberations on Findings/Recommendations; Report Discussion and Writing Status; Remaining Data Gathering Needed; Plans for September 16-20 Meeting (Woods Hole); Break-out into Subgroups (as needed).

Moderator: Dr. Annette J. Krygiel, Committee Chair

1500 ADJOURN

C

Acronyms and Abbreviations

ACC	Air Combat Command
ACTD	advanced concept technology demonstration
AC2ISRC	Aerospace Command and Control and Intelligence, Surveillance, and Reconnaissance Center
AFEO	Air Force Experimentation Office
AFOTEC	Air Force Operational Test and Evaluation Center
AFROC	Air Force Requirements Oversight Council
AJC2	Adaptive Joint Command and Control
ARCI	Advanced Rapid COTS Insertion program
ARG	amphibious ready group
ASN(RDA)	Assistant Secretary of the Navy for Research, Development and Acquisition
ASW	antisubmarine warfare
ATD	advanced technology demonstration
AWE	advanced warfighting experiment
CAOC-X	Combined Air Operations Center-Experimental
CBIRF	Chemical/Biological Incident Response Force
CDEC	Combat Development Experimentation Command
CDS	Combat Development System
CEC	cooperative engagement capability
CEE	Continuous Experimentation Environment
CENTCOM	U.S. Central Command
CFFC	Commander, Fleet Forces Command

CINC	Commander in Chief (now referred to as Combatant Commander)
CJCSI	Chairman of the Joint Chiefs of Staff Instruction
CJTF	Commander, Joint Task Force
CMC	Commandant, U.S. Marine Corps
CNO	Chief of Naval Operations
CONOPS	concept of operations
COP	common operating picture
COTS	commercial off-the-shelf
CPX	command post exercise
CROP	Common Relevant Operating Picture
CSG	carrier strike group
CTP	common tactical picture
CVBG	aircraft carrier battle group
CVG	carrier group
CW	Capable Warrior
CWL	Commandant's Warfighting Laboratory
C2	command and control
C3I	command, control, communications, and intelligence
C4I	command, control, communications, computers, and intelligence
C4ISR	command, control, communications, computers, intelligence, surveillance, and reconnaissance
DARPA	Defense Advanced Research Projects Agency
DDG	guided-missile destroyer
DFN	Digital Fires Network
DOD	Department of Defense
DOTMLPF	doctrine, organization, training, material, leadership, personnel, and facilities
DT&E	developmental test and evaluation
DVTE	Distributed Virtual Training Environment
EBO	effects-based operation
EFDS	Expeditionary Force Development System
ERGM	extended-range guided munition
ESG	expeditionary strike group
FAV	fast attack vehicle
FBCB2	Force XXI Battle Command Brigade and Below
FBE	Fleet Battle Experiment
FBE-I	Fleet Battle Experiment-India

FCS	Future Combat System
FOFAC	forward observer/forward air controller
FTX	field training exercise
FY	Fiscal Year
GCCS	Global Command and Control System
GCCS-M	Global Command and Control System-Maritime
GISRC	GCCS intelligence, surveillance, and reconnaissance capability
GPS	Global Positioning System
GSTF	Global Strike Task Force
HSV	high-speed vessel
HW	Hunter Warrior
IEEE	Institute of Electrical and Electronics Engineers
IFAV	Interim Fast Attack Vehicle
IKA	Information Knowledge Advantage
IO	information operations
IP	Internet Protocol
ISR	intelligence, surveillance, and reconnaissance
JBC	Joint Battle Center
JCATS	Joint Conflict and Tactical Simulation
JCDE	Joint Concept Development and Experimentation
JCS	Joint Chiefs of Staff
JEF	Joint Experimental Federation
JEFX	Joint Expeditionary Force Experiment
JFACC	Joint Force Air Component Commander
JFCOM	Joint Forces Command
JFE	Joint Fires Element
JI&I	joint interoperability and integration
JISR	Joint Intelligence Surveillance and Reconnaissance
JMCC	Joint and Maritime Component Commander
JMO-T	Joint Medical Operations-Telemedicine
JROC	Joint Requirements Oversight Council
JSAF	Joint Semi-Automated Forces
JSIMS	Joint Simulation System
JTAMDO	Joint Theater Air and Missile Defense Office
JTASC	Joint Training, Analysis, and Simulation Center
JTF	Joint Task Force
JVIMP	Joint Vision Implementation Master Plan

JWARS	Joint Warfare System
JWC	Joint Warfighting Center
JWCA	Joint Warfighting Capabilities Assessment
KB(X)	Kernel Blitz (Experimental)
LAWS	Land Attack Warfare System
LCAC	landing craft, air-cushioned
LELFAS	Long Endurance Low Frequency Active Sonar
LOE	limited-objective experiment
LTA	limited technical assessment
MAJCOM	Major Command
MBC	Maritime Battle Center (NWDC)
MC 02	Millennium Challenge '02
MC2A	Multi-Mission Command and Control Aircraft
MCCDC	Marine Corps Combat Development Command
MCM	mine countermeasure
MCMC	Mine Countermeasures Commander
MCP	Mission Capabilities Package
MCTSSA	Marine Corps Tactical Systems Support Activity
MCWL	Marine Corps Warfighting Laboratory
MEF	Marine Expeditionary Force
MEU	Marine Expeditionary Unit
MIO	Maritime Interception Operations
MIWC	Mine Warfare Commander
MOBA	Marine operations in built-up areas
MOP	Measure of Performance
MOUT	military operations in urban terrain
MROC	Marine Requirements Oversight Council
M&S	modeling and simulation
MTO	Maritime Tasking Order
NAVSEA	Naval Sea Systems Command
NBC	nuclear, biological, or chemical
NCASW	network-centric antisubmarine warfare
NCW	network-centric warfare
NETWARCOM	Navy Network Warfare Command
NFN	Naval Fires Network
NRL	Naval Research Laboratory
NSS	Naval Simulation System
NWDC	Navy Warfare Development Command

OC	Olympic Challenge
OD04	Olympic Dragon 2004
OEF	Operation Enduring Freedom
OMFTS	Operational Maneuver from the Sea
ONR	Office of Naval Research
OPNAV	Office of the Chief of Naval Operations
OPTEVFOR	Operational Test and Evaluation Force
OSD	Office of the Secretary of Defense
OTA	other transaction authority
PACOM	Pacific Command
POM	Program Objectives Memorandum
PTW	Precision Targeting Workstation
RDO	Rapid Decisive Operations
RISTA	reconnaissance, intelligence, surveillance, targeting, acquisition
SARGE	surveillance and reconnaissance ground equipment
SBCT	Stryker Brigade Combat Team
SJF HQ	standing joint force headquarters
SLOC	Sea Line of Communication
SLUAV	submarine-launched unmanned aerial vehicle
SME	subject matter expert
SOF	Special Operations Force
SPACECOM	U.S. Space Command
SPMAGTF	Special Purpose Marine Air/Ground Task Force
SSGN	cruise missile submarine
STOM	Ship to Objective Maneuver
SUBDEVRON	submarine development squadron
SYSCOM	Systems Command
T&E	test and evaluation
TAMD	theater air and missile defense
TBMD	theater ballistic missile defense
TCT	time critical target
TENCAP	Tactical Exploitation of National Capabilities
TES-N	Tactical Exploitation System-Navy
THS	Target Handoff System
TLDHS	Target Location and Designation Handoff System
TMD	theater misslie defense
TRADOC	Training and Doctrine Command

TTPs	tactics, techniques, and procedures
UAV	unmanned aerial vehicle
UCAV	uninhabited combat air vehicle
UGV	unmanned ground vehicle
USACOM	U.S. Atlantic Command
USAF	U.S. Air Force
USD	Under Secretary of Defense
USDA&T	Under Secretary of Defense for Acquisition and Technology
USLANTCOM	U.S. Atlantic Command (emphasis on fleet)
USJFCOM	U.S. Joint Forces Command
USMC	U.S. Marine Corps
USN	U.S. Navy
UW	Urban Warrior
VCNO	Vice Chief of Naval Operations
VLS	Vertical Launch System
WARNET	Wide-Area Relay Network
WIDT	Warfare Innovation Development Team
WRAP	Warfighting Rapid Acquisition Program

D

Definitions of Experimentation Terms Used in This Report

Advanced concept technology demonstration (ACTD)	One of three technology transition mechanisms; the other two are ATDs and experiments. ACTDs are used to determine the military utility of proven technology and to develop the concept of operations that will optimize effectiveness. ACTDs are not themselves acquisition programs but are designed to provide a residual, usable capability upon completion and/or transition into acquisition programs. Funding is programmed to support up to 2 years in the field. ACTDs are funded with Advanced Technology Development (ATD) funds. (Defense Acquisition University Glossary, 24 January 2003, http://deskbook.dau.mil/jsp/Glossary.jsp)
Advanced technology demonstration (ATD)	One of three technology transition mechanisms; the other two are ACTDs and experiments. ATDs are used to demonstrate the maturity and potential of advanced technologies for enhanced military operational capability or cost-effectiveness, and to reduce technical risks and uncertainties at the relatively low costs of informal processes. ATDs are funded with Advanced Technology Development (ATD) funds. (Defense Acquisition University Glossary, http://deskbook.dau.mil/jsp/Glossary.jsp)

Advanced warfighting experiment (AWE)	Large-scale warfighting experiments that explore emerging operational concepts and new technologies in an end-to-end manner. (Secretary of Defense William S. Cohen, Annual Report to the Congress, 1998, Chapter 15.) The term was used by the Army for its experiments, beginning in March 1997, designed to explore the brigade-level utility of digitization plans under its Force XXI program; it was later adopted by the Department of Defense, Congress, and others to include military experiments more generally.
Battlelab initiative	"Battlelab initiatives" originate with Service battle laboratories, and represent innovative or revolutionary concepts for operations or logistics to improve the capability to execute core Service competencies. They are expected to drive changes to organization, doctrine, training requirements, or acquisitions.
Command post exercise	An exercise in which the forces are simulated, involving the commander, the staff, and communications within and between headquarters. Also called CPX. (Joint Publication 1-02, Department of Defense Dictionary of Military and Associated Terms, 12 April 2001 (as amended through 25 September 2002)).
Developmental test and evaluation (DT&E)	1. Any testing used to assist in the development and maturation of products, product elements, or manufacturing or support processes. 2. Any engineering-type test used to verify status of technical progress, verify that design risks are minimized, substantiate achievement of contract technical performance, and certify readiness for initial operational testing. Development tests generally require instrumentation and measurements and are accomplished by engineers, technicians, or soldier operator-maintainer test personnel in a controlled environment to facilitate failure analysis. (Defense Acquisition University Glossary, http://deskbook.dau.mil/jsp/Glossary.jsp)
DOTMLPF	Doctrine, organization, training, materiel, leadership, people, and facilities. Source: Chairman of the Joint Chiefs of Staff Manual (CJCSM) 3500.03A, "Joint Training Manual for the Armed Forces of the United States," 1 September 2002, Glossary Part I – Abbreviations and Acronyms. Available at: http://www.dtic.mil/doctrine/jel/training_pubs/jtmmaster2002.pdf

Engineering experiment	An extension of scientific experiments to explore the technical problems, feasibility, and directions for application of the results of scientific advance to practical devices, machinery, and systems that are applicable to warfare.
Exercise	A military maneuver or simulated wartime operation involving planning, preparation, and execution. It is carried out for the purpose of training and evaluation. It may be a multinational, joint, or single-Service exercise, depending on participating organizations. (Joint Publication 1-02, Department of Defense Dictionary of Military and Associated Terms, 12 April 2001 (as amended through 25 September 2002)).
Experimentation campaign	An experimentation campaign is a planned and cohesive multi-year program of experimentation built upon a series of experiments and related activities to develop the knowledge needed to inform major decisions about future forces, explore the viability of potential or planned changes to forces or their capabilities, and/or confirm that planned capability development and directions will perform as expected.
Field exercise	An exercise conducted in the field under simulated war conditions in which troops and armament of one side are actually present, while those of the other side may be imaginary or in outline. (Joint Publication 1-02, Department of Defense Dictionary of Military and Associated Terms, 12 April 2001 (as amended through 25 September 2002)). As part of this definition, note that the fielded forces can also include units that simulate enemy systems and tactics to lend realism to the training exercise.
Field training exercise	An exercise in which actual forces are used to train commanders, staffs, and individual units in basic, intermediate, and advanced-level warfare skills. Also called FTX. (Joint Publication 1-02, Department of Defense Dictionary of Military and Associated Terms, 12 April 2001 (as amended through 25 September 2002)).
Fleet Battle Experiments (FBEs)	A series of Navy experiments begun in the mid-1990s to explore the implications of new technologies and concepts, including C4ISR, using fleet units and usually undertaken while the participating fleet units are engaged in training exercises. FBE Juliet took place during summer 2002.

Joint concept development and experimentation	A program sponsored by U.S. Joint Forces Command to explore and refine new concepts of joint warfare. These efforts are similar to Service-unique experimentation, but they involve more than one Service's forces and are typically planned, managed, and controlled by joint commands.
Joint Expeditionary Force Experiment (JEFX)	Series of Air Force experiments begun in 1998 (as EFX) to explore experimentally the implications of new technologies, organizations, and concepts, especially in the areas of C4ISR and logistics.
Joint experimentation	The application of scientific experimentation procedures to assess the effectiveness of proposed (hypothesized) joint warfighting concept elements to ascertain whether elements of a joint warfighting concept cause changes in military effectiveness.
Joint Warfighting Capabilities Assessment	A team of warfighting and functional area experts from the Joint Staff, unified commands, Services, Office of the Secretary of Defense, and Defense agencies tasked by the Joint Requirements Oversight Council with completing assessments and providing military recommendations to improve joint warfighting capabilities. Also called JWCA. (Joint Publication 1-02, Department of Defense Dictionary of Military and Associated Terms, 12 April 2001 (as amended through 25 September 2002)).
Joint warrior interoperability demonstrations	A series of biennial joint demonstrations coordinated by the Joint Staff to evaluate technologies from the private sector in a military environment, and to identify warfighter and C4I interoperability requirements and issues. The demonstrations are sponsored by a different combatant commander and led by a different Service every other year. (Derived from two sources: Joint Staff budget estimates for FY 2003; C4I for the Warrior, January 1998.)
Limited-objective experiment (LOE)	A relatively narrowly focused warfighting experiment that may be carried out by the Navy, Marines, or the U.S. Joint Forces Command, designed to explore a single issue or capability, not embedded in the fleet or in Marine units of significant size.

Military experiment	A military activity conducted to discover, test, demonstrate, or explore future military concepts, organizations, and equipment and the interplay among them, using a combination of actual, simulated, and surrogate forces and equipment. (Chapter 2).
Model	A representation of an actual or conceptual system that involves mathematics, logical expressions, or computer simulations that can be used to predict how the system might perform or survive under various conditions or in a range of hostile environments. (Defense Acquisition University Glossary, http://deskbook.dau.mil/jsp/Glossary.jsp)
Scientific experiment	Consists of the traditional problem statement, hypothesis, test, recording of results, conclusions, and recommendations for further work to advance the state of knowledge in a particular field of science.
Simulation	A method for implementing a model. It is the process of conducting experiments with a model for the purpose of understanding the behavior of the system modeled under selected conditions or of evaluating various strategies for the operation of the system within the limits imposed by developmental or operational criteria. Simulation may include the use of analog or digital devices, laboratory models, or "testbed" sites. Simulations are usually programmed for solution on a computer; however, in the broadest sense, military exercises, and war games are also simulations. (Defense Acquisition University Glossary, http://deskbook.dau.mil/jsp/Glossary.jsp)
Spiral development	"An iterative process for developing a defined set of capabilities within one increment. This process provides the opportunity for interaction between the user, tester, and developer. In this process, the requirements are refined through experimentation and risk management, there is continuous feedback, and the user is provided the best possible capability within the increment. Each increment may include a number of spirals. Spiral development implements evolutionary acquisition." Quoted from Memorandum from Under Secretary of Defense E.C. Aldridge, Jr., to the secretaries of the military departments and others, dated April 12, 2002.

Test and evaluation (T&E)	Process by which a system or components are exercised and results analyzed to provide performance-related information. The information has many uses including risk identification and risk mitigation and empirical data to validate models and simulations. T&E enables an assessment of the attainment of technical performance, specifications and system maturity to determine whether systems are operationally effective, suitable and survivable for intended use, and/or lethal. There are three distinct types of T&E defined in statute or regulation: Developmental (DT&E), Operational (OT&E), and Live Fire (LFT&E). (Defense Acquisition University Glossary, http://deskbook.dau.mil/jsp/Glossary.jsp)
War game	A simulation, by whatever means, of a military operation involving two or more opposing forces using rules, data, and procedures designed to depict an actual or assumed real life situation. (Joint Publication 1-02, Department of Defense Dictionary of Military and Associated Terms, 12 April 2001 (as amended through 25 September 2002)).